History In The Mapping

Four Centuries of Adirondack Cartography

Paul G. Bourcier

A Catalog of the Exhibition
June 15, 1984 - October 15, 1985

Adirondack Museum
Blue Mountain Lake, New York

Front Cover: Section of *A New and Accurate Map of the present War in North America*, Richard William Seale, 1757.

Back Cover: "North Creek," from *County Atlas of Warren, N.Y.*, Frederick W. Beers, 1876.

ISBN 0-910020-37X
Library of Congress Catalog Card Number: 86-70110
Manufactured in the United States of America
Design by Jane Mackintosh

Preface and Acknowledgements

A map, like any artifact, is a product of the historical environment in which it is made. The nature of this environment may be ascertained by studying maps in two ways: first, by examining the reason for a map's existence and determining its context, social, cultural, economic or political; and second, by studying the information conveyed on the map itself. The history of a region's mapping and the maps of a region's history together reveal intriguing insights about the past.

Maps of the Adirondack area of northern New York suggest three stages in the history of the region. Referred to in this catalogue as "Unknown Wasteland," "Advancing Frontier," and "Wilderness Resort," these dramatize man's changing relationships to the Adirondacks during the past four centuries.

The exhibition, *History in the Mapping*, was planned with these broad themes in mind. The exhibit featured sixty-six maps and atlases from the collections of the Adirondack Museum Library, a repository of over two thousand cartographic items of regional interest. Special thanks are extended to the museum librarian, Jerold Pepper, for his cooperation.

The research for the exhibition and the catalog essay benefitted substantially from the assistance of the staffs at the New York State Library in Albany and the New York Public Library in New York City. The contributions of Adirondack historian Warder Cadbury are also greatly appreciated.

The exhibition would not have been possible without the support of the Adirondack Historical Association and the guidance of William Crowley, former curator at the Adirondack Museum. The author is indebted to the museum's director, Craig Gilborn, for the time and talent spent in reading and improving the manuscript for this publication.

This catalogue sketches the history of Adirondack cartography from the seventeenth century to today, and is intended as a brief overview of the ways maps have reflected and contributed to the development of New York's northern wilderness. It discusses some of the most significant maps of the region as well as maps which were most representative of a theme or trend.

Numbers noted in the essay indicate the checklist numbers of exhibited maps. These maps are listed in numerical order at the back of this publication.

We are pleased to be able to bring such a significant map collection to public view and trust that it contributes to greater understanding of the history of the Adirondack region.

The Unknown Wasteland

When, in 1609, Samuel de Champlain first sailed the lake which now bears his name, European traders and explorers were unlocking the geographical secrets of a new continent. Traveling largely upon rivers and lakes, they charted the major valleys of the New World without venturing far into the isolated interior regions between their routes of travel. Champlain had no real conception at all of the area now known as the Adirondacks, which is generally bounded by the valleys of the St. Lawrence and Mohawk Rivers to the north and south, and Lakes Champlain and Ontario to the east and west. Champlain's *Carte de la nouvelle france. . .* of 1632 illustrated principal river valleys with perspective drawings of Indian habitations and mountains, but without definite locations. Because he knew nothing about the vast Adirondack interior, Champlain allocated virtually no space between major rivers and lakes [#2]. Lack of familiarity with inland territories also contributed to grossly inaccurate renderings of the relative positions of known river valleys; early French and Dutch maps placed Lake Champlain where New Hampshire is today.

The Adirondacks began to assume an obscurely conceived identity on maps by the late 1600s. The Dutchman Nicolao Joannis Visscher II labeled the amorphous region between the Mohawk and St. Lawrence Rivers as New Belgium on his 1685 map, *Novi Belgii, Novaeque Angliae nec non partis Virginae Tabula. . .*[#3]. The Dutch, who settled along the Hudson Valley and involved themselves deeply in a fur trade with native Americans, vaguely imagined "Nova Belgica" as a mountainous, forested region rich in fur-bearing animals. Hence, Visscher illustrated the area with perspective drawings of trees, mountains and wild creatures without exact locales.

By the early eighteenth century, the French began to leave space on their maps for today's Adirondacks, then known only as lands assigned to the Iroquois. The most important cartographic contributions of the French, however, were progressively detailed renderings of "Lac Champlain" and "Lac du Sacrement" (Lake George), evident on the maps of Guillaume de L'Isle. Royally appointed French surveys of the Champlain Valley yielded information that was reasonably accurate about the geography of the eastern border of the Adirondack

I. du Borgue
Uttawas or Outawacs R.
R. Rouge
F. Sorrel
MONTREAL Is.
Guilleson
WACS
the Iroquois
Rideau R.
Montreal
La
Longville
E. Assomption
Bonne Chere R.
Mataouachic R.
IROQUOIS
The Ferry
St. Louis
Prairie
Iroquois or
L. St. Francis
Extent of the French Settlements, before they built a Fort at Crown Point.
F. Ch
ERN
Long Fall
Battoe I.
Parts but little known
Rapide
Catarakui R.
ANTIENT
Iroquois or Champlain Lake
la Gallette
Presentation P.
I. la Motte
Amand R.
of
Toinata
or
Chouagatchi R.
No. Branch of Hudsons R.
IROQ
Iroquois
Great I.
NEW
Niouure Bay
Crown Point
Assumption R.
YORK
Lake George
Wood Creek
Gen. Johnsons Camp
Eamine R.
DES
Burnets Fields
W. Branch of Hudson R.
Palatines T.
F. Edward
French
Oneida L.
F. Ann
F. Nicholson
Mohoks R.
MOHOCKS
Saraktoga
ONEIDES
Hudsons R.
G. Johnsons Seat
F. Hunter
ES
Onondaga
Tuscarros C.
Mohoks Cas
Gaientoha L.
Susquehanna E. B.
Housuk R.
ONDAGES
Sheneklady F.
Senang or Seneca
TUSCARROS
Germans T.

region.

In contrast, cartographic depiction of the interior Adirondacks would for decades be conjectural, based on hearsay and supposition. Its rugged terrain and isolated geography made the region a cartographic backwater during the colonial period. European traders and soldiers had little reason to attempt to explore the unknown wilderness and dismissed the area as useless. British mapmakers invented an imaginary topography as if to compensate for their lack of knowledge. In 1733, the British cartographer Henry Popple filled the Adirondack area with drawings of peaks and trees in *A Map of the British Empire in America with the French and Spanish Settlements adjacent thereto*. Popple also extended known rivers like the Hudson along extrapolated paths to imaginary sources, with results that were far from accurate.

English mapmakers frequently used conjectural drawings to portray unknown territories, thereby making hypothesis appear to be fact. On his 1755 *Map of the British Colonies in North America. . .*, John Mitchell demonstrated his vague conception of the Adirondacks by drawing mountains across the eastern portion of the region and by writing "Marshes and Mountains" across the western portion. The London engraver and draftsman Richard William Seale included speculative notions on his *New and Accurate Map of the Present War in North America*, published in 1757 [#5]. He extended the courses of the Hudson and Oswegatchie Rivers along fabricated avenues into "parts but little known."

Situated on the doorstep of French Canada, the Champlain Valley became a strategic corridor in the armed struggle between France and Britain for control of New England and northern New York. Both sides conducted military surveys of the valley and produced detailed, large-scale maps of various localities within it. Forts St. Frederic (Crown Point), La Reine (Fort Ann), Carillon (Fort Ticonderoga), William Henry (Fort George), and Edward, constructed on the western side of the Champlain Valley, were the subjects of these detailed cartographic plans. Military engagements were chronicled in maps such as *A Perspective View of the Battle fought near Lake George on the 8th of Sept. 1755. . .*, drawn in 1756 by Samuel Blodget [#4].

By 1750, France and Britain had gained an intimate familiarity with the islands, bays, river mouths and true positions of Lake

---

Section of *A New and Accurate Map of the present War in North America*, Richard William Seale, 1757. This map demonstrates a familiarity with the militarily-strategic Hudson-Champlain corridor, but it labels the Adirondacks as "Parts but little known." Rivers in this area have conjectural courses.

Champlain and Lake George. William Furness Brassier and other cartographers delineated the Champlain shoreline, including the outlets of the lake's tributary rivers (the Chazy, the Saranac, the AuSable, the Boquet and others), but since their surveys were limited to the lake itself, their geographic knowledge ended a few miles inland. Detailed and accurate renderings of the Champlain Valley stood in stark contrast to the conjecture and omission that characterized the mapping of the interior Adirondacks.

This gulf between the known fringe and the unknown interior persisted throughout most of the eighteenth century, as mapmakers turned to general verbal descriptions for shreds of evidence. Geographer Lewis Evans labeled the core region "Couxsaxrage," an Indian term said to mean "dismal wilderness" or "beaver hunting ground." In 1755, he noted on *General Map of the Middle British Colonies in North America...* that "...this Country by reason of Mountains, Swamps and Drowned Land is impassible and uninhabited." Not surprisingly, Evans' depiction of the area was a nullity — a large blank space except for a vague range of mountains. Major Samuel Holland, following Evans' example, represented the Adirondack interior as a void on his 1776 map, *The Provinces of New York and New Jersey...*[#7]. Holland described the region,

> "Coughsaghrage, the Beaver Hunting Country of the Confederate Indians...It belongs to New York and is full of Swamps, Lakes, Rivers and Drowned Lands; a long chain of Snowy Mountains which are seen from Lake Champlain runs thro the whole Tract North and South. This Country is not only uninhabited but even Unknown...."

That same year, Lewis Evans' friend, the noted colonial administrator and political theorist Thomas Pownall, also evinced the disparity of the Adirondack/Champlain portrayal: Pownall's *A Map of the Middle British Colonies in North America* placed a legend of many detailed features of the Champlain Valley in the place left blank for the Adirondacks [#6].

Victory in the French and Indian War in 1763 gave the British control of the Champlain Valley and allowed the colony of New York to begin developing its northern territory. People began to perceive

---

Section of *A Map of the Province of New-York...*, Claude Joseph Sauthier, 1776. This map illustrates the contribution that the Totten and Crossfield Purchase of 1772 made to the cartography of the interior Adirondacks. While most of the region is devoid of any information, the Purchase is shown as a detailed watershed bridging Tryon and Charlotte Counties.

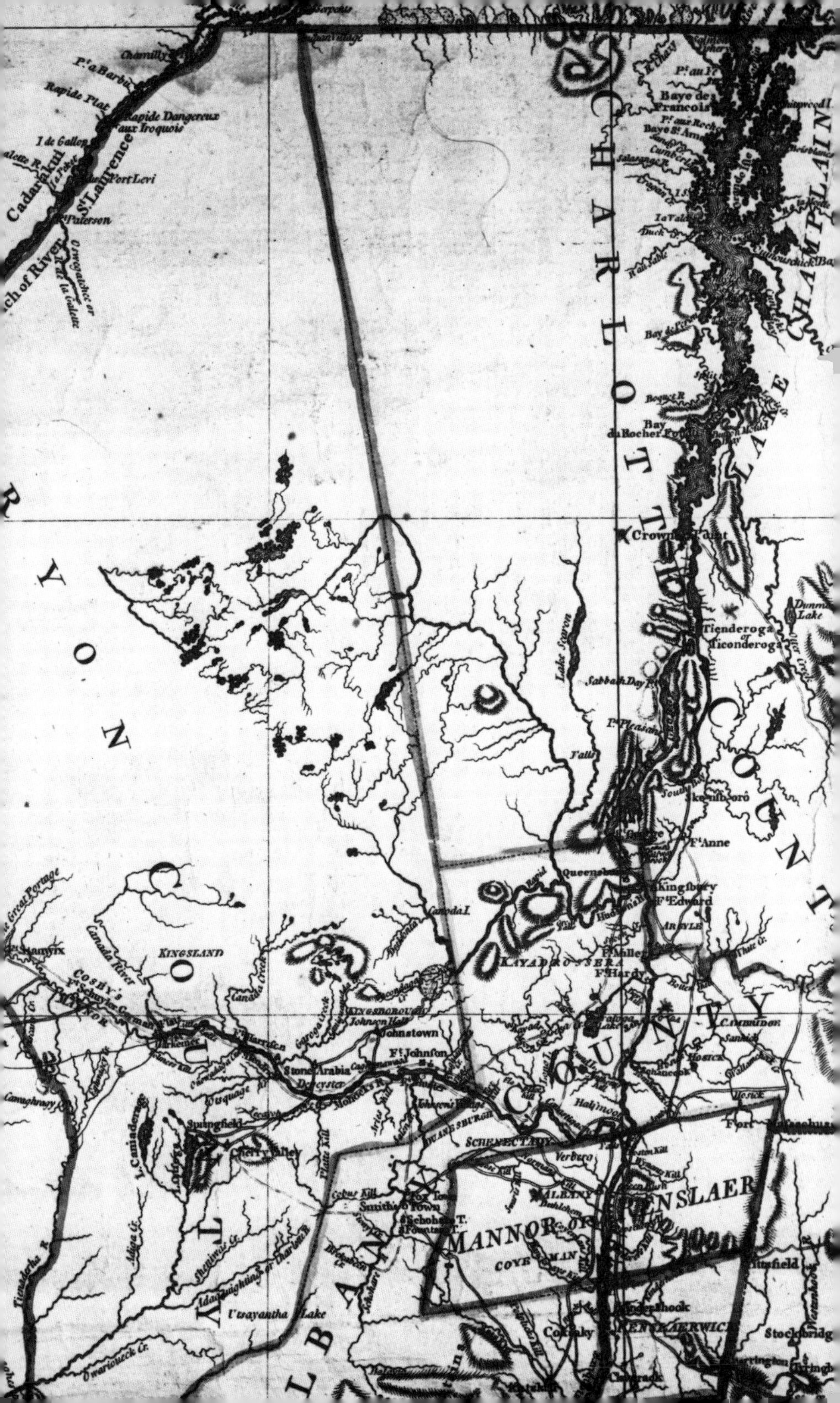

CHARLOTTE COUNTY
LAKE CHAMPLAIN
TRYON COUNTY
ALBANY COUNTY
MANNOR OF RENSLAER
Crown Point
Ticonderoga
Johnstown
Johnson Hall
Fort Johnson
Stone Arabia
Cherry Valley
Springfield
Schenectady
Albany
Kinderhook
Queensbury
Kingsbury
Ft. Edward
Ft. Anne
Skenesboro
Cambridge
Hosick
Saratoga
Kingsland
Canada Creek
Canada River
Rapide Dangereux
aux Iroquois
Port Levi
St. Lawrence
Cadarakui
Rapide Plat
Pt. a Barbu
Chamilly
Lake Scaron
Sabbath Day Pt.
Kayadrossera
Utsayantha Lake
Stockbridg
Pittsfield
Dunmore Lake
Baye de Francois

the Adirondacks as a region with economic potential for cultivation and settlement. This view of the area and the opportunity to act upon it marked the beginning of the end of the era of the "Unknown Wasteland."

It was the Totten and Crossfield Purchase of 1772 which first opened the interior to geographic scrutiny. Acting on behalf of a party of colonial investors, Joseph Totten and Stephen Crossfield petitioned the British governor of New York in 1771 for a license to extinguish Indian title to a large wedge-shaped tract of land comprised of 1,150,000 acres in the southeastern portion of the Adirondack plateau. Although the brothers Edward and Ebenezer Jessup were the motivating forces behind the purchase, the land transaction came to be known as the Totten and Crossfield Purchase. Interested in developing the territory, Ebenezer Jessup conducted a survey of the area in 1772, dividing the tract into fifty square "townships" arranged in a grid pattern on a northeast/southwest axis. Ebenezer's survey, appearing on a map entitled *A Plan of the Lands Purchased for the Benefit of Joesph Totten & Stephen Crossfield. . .*, yielded the first detailed information concerning a portion of the Adiriondack interior [#8]. Although Jessup neglected to label most of what he delineated, map readers were able to note for the first time many natural features, Raquette Lake and Indian Lake among them. Conjecture had been replaced at last by a modicum of truth.

Working in the wilds with compasses and tree markers, Jessup and his team of surveyors were hardly infallible. A number of cartographic errors appeared on the Jessup map of 1772, the most obvious one concerning the course of the Hudson River. Jessup mistook the headwaters of the then unknown Raquette River for the source of the Hudson. Since the Adirondack region outside the Totten and Crossfield Purchase remained unsurveyed, Jessup incorrectly connected the northeasterly flowing waters of modern-day Raquette, Forked and Long Lakes with the southeasterly course of the true Hudson River. The error would be corrected by those directly associated with the venture by 1776.

Cartographers soon borrowed from the Jessup survey. Four years after Jessup's map, the Purchase appeared conspicuously on *A Map of the Province of New York. . .* by topographer Claude Joseph Sauthier [#9]. Sauthier filled the Adirondack void with an area of detailed geographic information confined within the obvious wedge-shaped boundary of the lands that Jessup had surveyed, including Jessup's misconception that the Hudson River flowed from Raquette Lake.

The American Revolution postponed attempts to develop the newly surveyed areas of the Adirondacks. Some of the Loyalist purchasers of the Totten and Crossfield lands fled to Canada during the war, and after the conflict ended in 1783, the State of New York assumed dominion over this territory and redistributed its townships, bringing to a close a period marked by geographic ignorance and by military and trade interests confined to the Champlain Valley.

The Advancing Frontier

With the end of hostilities, a variety of people – veterans, farmers and their families, immigrants and speculators – turned their attention toward New York's northern wilderness. Inhospitable though it may have been, it was presumed to be rich in natural resources.

The Natural Landscape, 1784-1834

When Simeon DeWitt became surveyor-general of New York in 1784, his primary mission was the economic welfare of the state. To carry out a system of internal improvements, DeWitt encouraged private surveys, particularly of areas of New York still relatively unknown. Surveyors penetrated many areas of the Adirondack interior, mapping a variety of land grants and purchases for men who saw speculative opportunity in the state's northern wilderness.

Among the territories sold to private speculators by the state was Macomb's Purchase, covering more than three million acres in the northern and western Adirondacks. The transaction was recorded in *A Map of A Tract of Land in the State of New York called Macomb's Purchase*, by Charles C. Brodhead, who surveyed the area in 1792 [#10].

Because water provided transportation and power essential to economic development, maps fairly portrayed North Country waters as early as 1802. That year, Simeon DeWitt compiled *A Map of the State of New York* on which he delineated the rivers and lakes of the Adirondack region with a detail and accuracy never before achieved; past misconceptions were absent, and all but the smallest and most remote hydrological features were included.

In the years following the 1802 DeWitt map, mapmakers began to label many of the rivers and lakes of the central Adirondacks. The most important cartographer in this respect was John H. Eddy of New York City, who in 1818 compiled a map called *The State of New York with part of the adjacent states* [#12]. Eddy named many remote features, such as "The Long Lakes" (comprising what is now Raquette, Forked and Long Lakes), Lake Pleasant, and Lake Placid. His map also marked a step forward by using the planimetric method of hachuring to indicate individual mountain peaks; this technique replaced perspective drawings shown on earlier maps.

TOTTEN & CROSFIELDS
PURCHASE
A wild barren tract extends hereabouts, the property of the State, covered with almost impenetrable Bogs Marshes & Ponds, and the uplands with Rocks & evergreens.
Browns
Tract
Lake George
Long Lake
Elizabeth Town
Crown Point
Scroon
Minerva
Chester
Johnsbury
Athol
Hague
Bolton
Caldwell
Queensbury
Kingsbury
Fort Ann
Luzerne
Boonville
Remsen
Western
Steuben
Lake Pleasant
Round L.
Piseka Lake
Oxbow L.
Sacondago River
Moose River
Beaver River
Black River
Fish L.
High Falls 63 Ft.
Cranberry L.
Head of Hudsons R.
Head of Racket R.
Head of Moose R.
State Road
State Road from Russel to the Fish House
from Chester to Russell
Intended road
Glenns Falls
Fort Edward
Sandy Hill
Ticonderoga
Alexandria
Rogers Rock
Natural Bridge
Friends Lake
North or main Branch of Hudson
Jessups River
Jessups Falls
Crane Mt.

Although hachuring became accepted practice by the 1820s, few Adirondack mountains were delineated in detail due to their inaccessibility. People, cartographers among them, believed that New York's loftiest mountains were in the Catskills until as late as 1837.

The Human Landscape, 1784-1840

Maps from this period evidence three important elements of development: a system of roads, early industries, and pioneer settlements in the region's periphery.

In the two decades following Simeon DeWitt's appointment as surveyor-general of New York in 1784, an influx of transplanted New Englanders established farming communities in the Champlain and St. Lawrence Valleys. By 1804, cartographers were mapping villages which ringed the Adirondacks; these included Plattsburgh, Willsborough, Bolton, Ogdensburg, Lisbon and Massena.

Settlement brought a network of roads between new communities. Roads were restricted to the lowlands surrounding the Adirondack plateau, except for the so-called "military roads" which Amos Lay depicted on his 1812 *Map of the Northern Part of the State of New York* [#11]. These crude routes, which severally extended from Russell to the "Fish House" near Northville, from Russell to Chester, and from Hopkinton to Northwest Bay (Westport), ran in a northwest-southeast direction and were the first to penetrate the Adirondack interior. The State constructed the roads about 1810 in order to link commercial and administrative settlements in the St. Lawrence Valley with Albany and the Hudson-Champlain corridor. The misnomer "military road" was later applied to these routes due to an unfounded legend that soldiers had built them during the War of 1812. Most segments of the three "military roads" fell into disuse and disrepair by the time of the Civil War.

Small-scale local industry played a crucial role in the growth of the farming villages of the Adirondack periphery. By and large, cartographers and map publishers tended not to portray the manufactories of this area until the iron industry began to develop in the 1810s and 1820s. John H. Eddy, Henry Schenck Tanner of Philadelphia, Fielding Lucas, Jr. of Baltimore, and other mapmakers

---

Section of *The State of New York with part of the adjacent states*, John H. Eddy, 1818. On this map, the Adirondacks are seen as a territory on which land speculators and early surveyors have left their mark. The region is divided into tracts and townships, its hydrography is fairly complete, and three state roads penetrate the wilderness.

who chronicled New York's progress took note of the proliferation of forges and ironworks in the shadow of the still unsurveyed high peaks area of Essex County. The iron industry of the Champlain Valley was made possible by large deposits of ore, limestone and charcoal, together with abundant water power and good transportation. Completion of the Champlain Canal in 1823 further spurred the industry, since the canal connected Lake Champlain to the Hudson and Mohawk Rivers, thereby providing manufacturers with an all-water route to major markets to the south. Canal and river transportation allowed New York State to tap the potential of its northern frontier and fostered the growth of Plattsburgh, Port Henry, and other trading "ports" along Lake Champlain.

At this time, in the 1820s and 1830s, mapmakers of New York focused primarily on transportation routes important to early commerce and tourism. *The Tourist's Map of the State of New York*, published in 1827 by William Williams of Utica, typified these commercially produced state maps [#13]. Showing profiles of the newly completed Erie and Champlain Canals, and including a stage, canal and steamboat register, the Williams map was clearly a product of the drive for internal improvement. Several years later, when transportation began to shift to the railroad, maps of New York began to focus on railways as well.

The most significant maps of the century's second quarter were made by the noted geographer, David H. Burr. An act passed by the legislature in 1827 provided for the publication of a map and atlas of New York compiled by Burr and supervised by surveyor-general Simeon DeWitt. Completed in 1830 (although dated 1829), Burr's *Atlas of the State of New York* featured a map of every county in the state at a scale of one inch to two-and-one-half miles; it was the first to show in detail the extent and nature of settlement, transportation and industry in the Adirondacks [#28]. According to Burr, settlement in the region followed the ring pattern evident in less detail on earlier maps. Most towns and villages were on or near major lakes and rivers to take advantage of water power and transportation. Cartographic symbols of small industries and public buildings indicated the activities of these settlements: flour mills, manufactories, forges, furnaces, ore beds, saw mills and churches dotted the landscape. A network of stage roads and county roads stretched out along the

Section of *Map of the County of Essex*, David H. Burr, 1839. A major theme of this map is the growth of industry and settlement in the Adirondack periphery. The tiny symbols drawn along major rivers indicate the locations of flour mills, manufactories, forges, furnaces, saw mills and churches.

Rattlesnake P.
Montresor
Ross
WILLSBOROUGH
Willsborough P.O.
Field
Benson
Boquet River
Wriesburgh
Connelly
Slackville
Hicks
Essex
Bouquet
Wessex P.O.
Potts
Rogers
LEWIS
Lewis
Rogers
Wharton
Whallons Mills
ESSEX
Freswell
Judd
Platt Rogers
Wadhams
Elizabethtown P.O.
Ore
Morgan
WESTPORT
North West Bay
Shaw
ELIZABETHTOWN
Morgan
Williams
Wooley
Black P.
Long P.
Ore
Ore
Bossborough
Head Tract
Skene

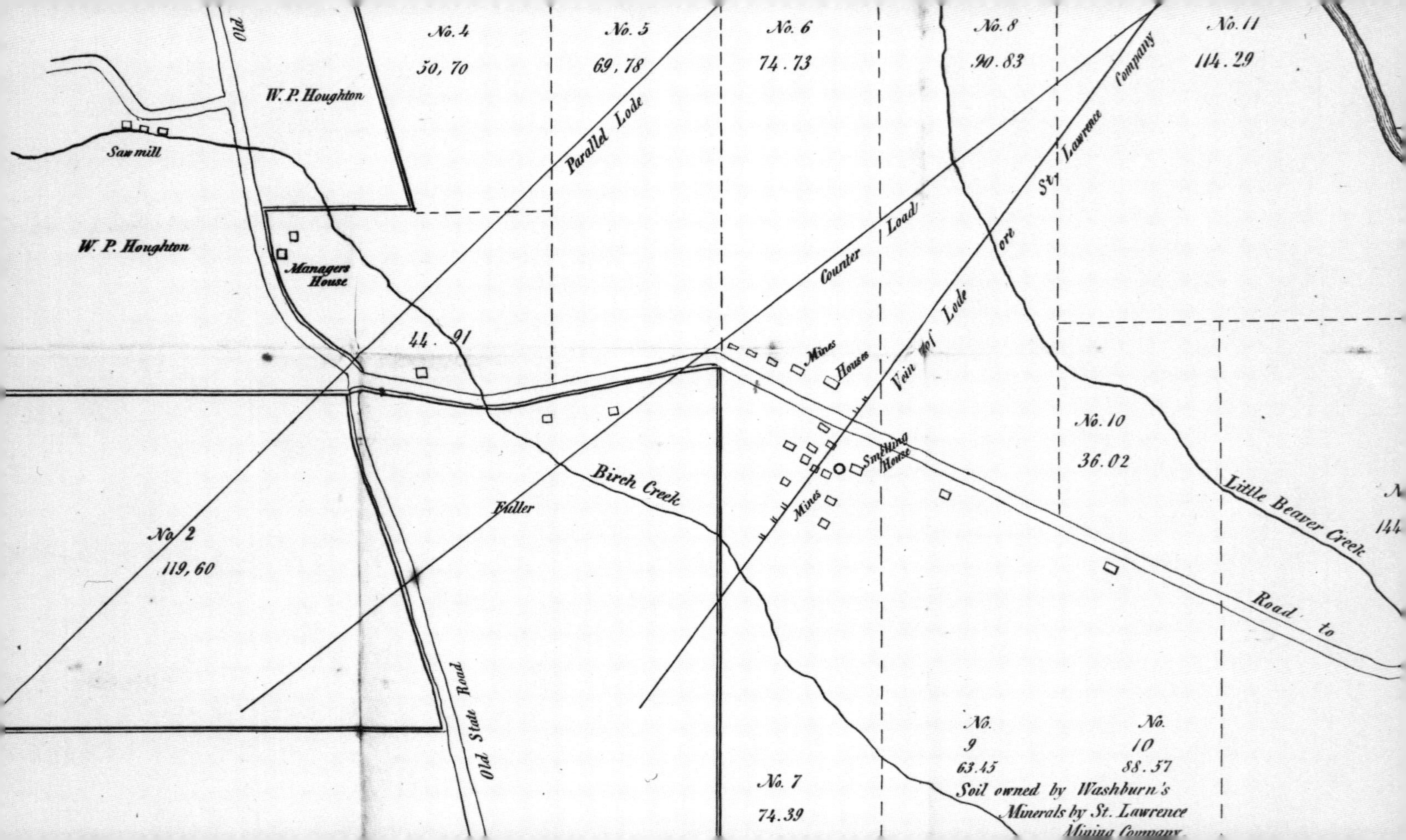

No. 4
50, 70
No. 5
69, 78
No. 6
74.73
No. 8
90.83
No. 11
114.29
Old
W. P. Houghton
Saw mill
W. P. Houghton
Managers House
Parallel Lode
Counter Load
St. Lawrence Company
ore
Vein of Lode
44. 91
Mines
Houses
Smelting House
Mines
No. 10
36.02
Little Beaver Creek
144
Birch Creek
Fuller
No. 2
119, 60
Road to
Old State Road
No. 7
74.39
No.
9
63.45
No.
10
88.57
Soil owned by Washburn's
Minerals by St. Lawrence
Mining Company

lowland valleys, linking villages one to another.

After publication of his first atlas and wall map, Burr continued to make and revise his works through the 1830s. Subsequent editions, published by Joseph Hutchins Colton of New York City, reflected the accelerating pace of development in the North Country. Adirondack counties depicted in Burr's atlases of 1839 and 1841 were far different than they had been just ten years earlier [#29]. The frontier itself was slowly moving inward toward the Adirondack core, as small mill and farming communities were established farther upstream along the valleys of the Oswegatchie, Grass, St. Regis, Chazy, Chateaugay, AuSable, Hudson, and Sacandaga Rivers. Political and property boundaries were altered as the population grew, and knowledge of the countryside increased with the aid of private surveys by logging interests and land developers.

Scientists, Engineers and Capitalists, 1836-1884

The State of New York began its Natural History Survey in 1836 to investigate the economic potential of its agricultural and mineral resources. Professor Ebenezer Emmons conducted the geological portion of the survey for the Adirondack region during the summers of 1836 through 1840. A landmark in the history of the Adirondacks, Emmons' survey profoundly affected the mapping of the area, ushering in an era of scientific study in the North Country.

Emmons made significant strides in surveying the more remote and least known areas of the Adirondack heartland. Perhaps the most noteworthy of his efforts was his barometric measurement of the altitudes of the chain of peaks he christened "The Adirondack Mountains." Cartographers such as David Burr and John Calvin Smith quickly took note of the new information generated by the Emmons survey and modified their maps accordingly.

The first maps of the geology of the region appeared in reports that Emmons submitted to the governor in the late 1830s and early 1840s. An example was a *Geological Map of Clinton County*, printed in 1842 [#19]; characteristic of the new scientific generation of maps, it outlined with color coding a pattern of geologic formations which underlay the surface features that other mapmakers were charting. Emmons and his associates defined the Adirondack re-

---

Section of *Map of. . .Lands belonging to the St. Lawrence Mining Co. . . .*, 1852. This map is typical of maps made for the owners of Adirondack mining operations. It shows the property of a lead mining business near Pierce's Corners, N.Y. Important details, such as the locations of ore deposits and company-owned structures, are prominently illustrated.

gion as a distinct zone, labeled "primary system" on the 1842 *Geological Map of the State of New York, by Legislative Authority* [#18]. Emmons viewed the Adirondacks as a well-defined agricultural region as well, an area he noted as having "but little arable soil" on his 1846 *Agricultural Map of the State of New-York* [#20].

Although the Natural History Survey discouraged the prospect of commercial farming in the area, it publicized the region's potential for extractive industry in its glowing reports of vast deposits of iron ore and other minerals. The subsequent growth of Adirondack mining led to the production of numerous maps showing the physical plants of mining enterprises. One of the very first of these appeared in an Emmons report of the late 1830s; titled *Mining District of Rossie*, it indicated veins of lead ore, an ironworks, and a copper mine in western St. Lawrence County. Later mining maps, such as the 1852 *Map of. . .Lands belonging to the St. Lawrence Mining Co. . . .*, featured vertical profiles illustrating cross-sections of ore deposits and mine shafts [#23].

By mid-century, the Adirondack Iron District of western Essex County embraced the most publicized and most mapped mining area in the Adirondacks. Home of the highly celebrated McIntyre Mine, this district captured the attention of Emmons and other geologists and surveyors. In the 1840s and 1850s, E. N. Horsford and other cartographers made several maps of the area which were used by local entrepreneurs as promotional tools to encourage investment in the iron industry. Among these maps were Horsford's *Map exhibiting the relative position of the Veins of Ore in and adjoining the Village of McIntyre. . .* and *Map of the County of Essex*. In 1879, E. V. D'Invilliers made a *Geological Map of Essex County*, noting the extent of various types of iron ore and the location of mining villages in the midst of this rich mineral region [#22]. This map demonstrated the continuing benefits of scientists to the Adirondack mining industry.

Owners of local mines also hired surveyors to determine the extent of their properties and mineral rights. To this end, D. M. Arnold of Ticonderoga made a *Map of Lands and Ore Rights in Lots 20, 21, 22, 23, 24 & 25 Iron Ore Tract Essex County N.Y.* for the Port Henry Iron Ore Company in 1865; also, a *Map of a part of Hochstrasser Lot Dannemora N.Y.* ten years later for Clinton State

Section of *Geological Map of Essex County, New York*, E.V. D'Invilliers, 1879. The shaded areas on this map represent different geological formations which underlay a mineral-rich area of the Adirondacks. The letters "A," "B" and "D" indicate the locations of iron ore deposits. Mining villages such as Irondale and Mineville are also noted.

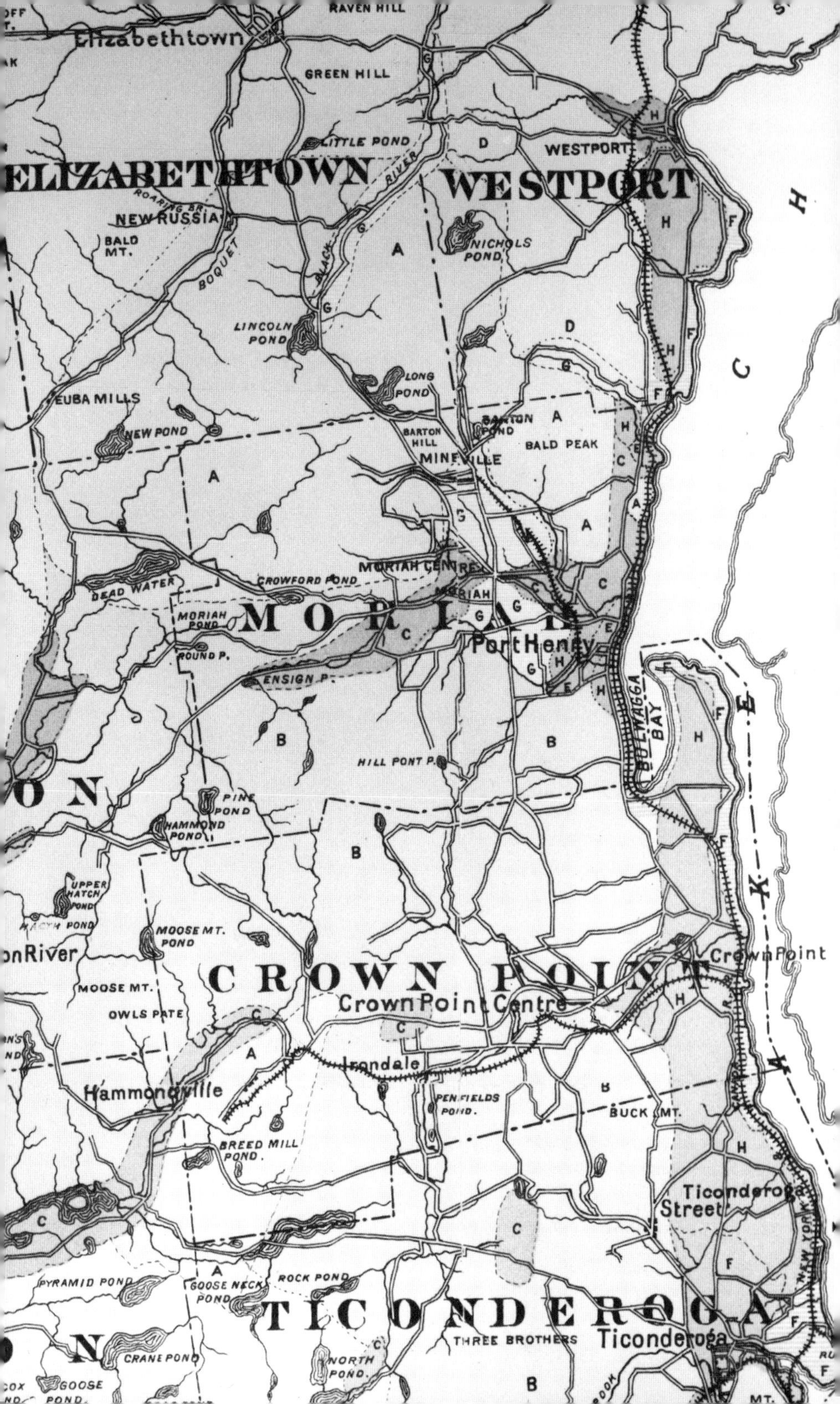

RAVEN HILL
Elizabethtown
GREEN HILL
LITTLE POND
ELIZABETHTOWN
WESTPORT
WESTPORT
ROARING BR.
NEW RUSSIA
BALD MT.
BOQUET
BLACK
RIVER
NICHOLS POND
LINCOLN POND
LONG POND
EUBA MILLS
NEW POND
BARTON HILL
BARTON POND
BALD PEAK
MINEVILLE
MORIAH CENTRE
CROWFORD POND
DEAD WATER
MORIAH
MORIAH POND
MORIAH
ROUND P.
Port Henry
ENSIGN P.
BULWAGGA BAY
HILL PONT P.
PINE POND
HAMMOND POND
UPPER HATCH POND
MOOSE MT. POND
CROWN POINT
Crown Point
Crown Point Centre
MOOSE MT.
OWLS PATE
Irondale
Hammondville
PENFIELDS POND
BUCK MT.
BREED MILL POND
Ticonderoga Street
PYRAMID POND
GOOSE NECK POND
ROCK POND
TICONDEROGA
THREE BROTHERS
Ticonderoga
CRANE POND
NORTH POND
GOOSE POND

Prison, a state-run iron mining and smelting operation worked by convicts [#24]. Since these maps were used only by the mine operators, they were manuscript productions, never lithographed for printing.

Scientists noted not only the extensive mineral resources of the Adirondacks, but also the commercial potential of the region's vast timber stands. Lumbermen were conducting annual river drives, sending softwood logs downstream to be cut at the many sawmills which lined the region's rivers. Colliers were harvesting great tracts of hardwood to make charcoal for the forges and furnaces of the local iron industry. Others fed hemlock bark to local tanneries and timber to the area's potash makers.

Surveyors almost always preceded loggers into wilderness tracts, mapping out available timber stands and establishing property lines. For this purpose, W. E. & J. Sibell of New York City published a *Map showing the location of. . .Pine Timber. . .* in 1854 for the Adirondack Iron and Steel Company, and R. Fairbanks drew a map of the "Ord" tract of the Hudson highlands in 1855 for Cheney and Arms, a Glens Falls lumber company [#25 and #26].

The exploitation of natural resources in the Adirondacks depended on effective delivery of raw materials to market. Engineers and businessmen issued a number of proposals for transportation systems designed to tap the region's timber and mineral wealth. Land speculators encouraged these proposals to increase the accessibility and value of their personal holdings.

In 1838, the State of New York authorized a survey of several routes for a railroad to cross the northern Adirondacks between Ogdensburg and Lake Champlain. Edward H. Brodhead, chief engineer of the project, and his assistant, Professor Edwin F. Johnson, surveyed the proposed routes – the northernmost following the lowland border of the region through Malone, the southernmost actually penetrating the wilderness interior through the valleys of the lower Raquette, the upper Saranac, and the lower AuSable Rivers [#14]. In compiling their maps of the proposed routes, Brodhead and Johnson made certain that they included information about the locations of the major mineral deposits adjoining the rights-of-way.

Section of *Untitled* [*Survey of a Railroad and Steamboat Route from Lake Champlain to the County of Oneida*], Farrand N. Benedict, 1846. The heavy, dotted line on this map shows a portion of a proposed commercial steamboat route designed to tap the resources of the interior Adirondacks. Like many attempts to penetrate this rugged, isolated region, this venture met with failure.

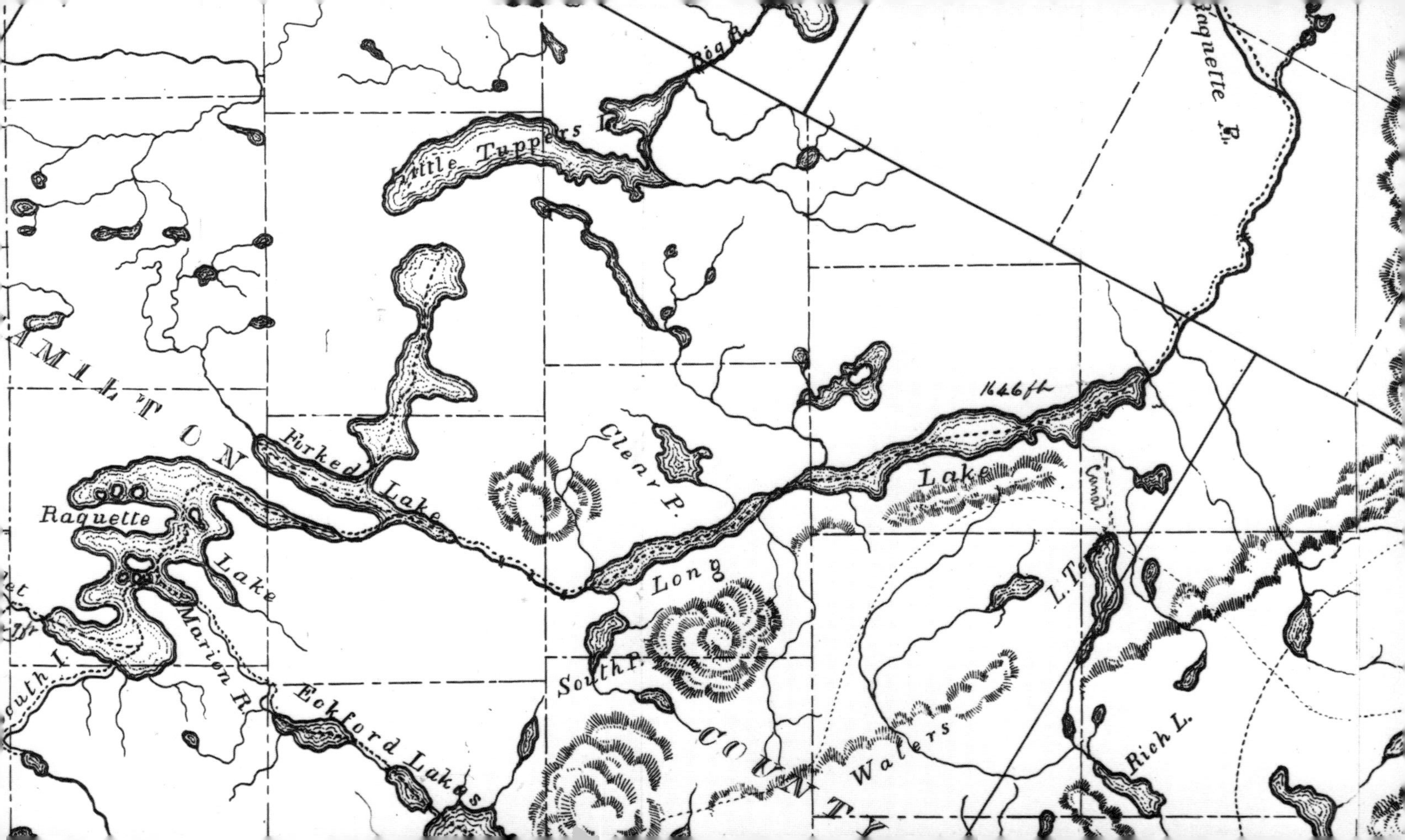

Raquette R.
Little Tuppers L.
Forked Lake
Clear P.
1646 ft
Long Lake
Canal
L. Ter.
Rich L.
Waters
South P.
COUNTY
Raquette Lake
Marion R.
Eckford Lakes
AMILTON

Investors eventually decided to build along the more easily traversed northern route, which was completed in 1850.

One of the most ambitious but ill-fated attempts to tap New York's northern heartland began with the observations of Professor Farrand Northrup Benedict, a colleague of Ebenezer Emmons. Benedict, a civil engineer and mathemetician, studied the topography of the Adirondack plateau during the late 1830s and 1840s. He noted as early as 1835 the existence of a chain of navigable lakes penetrating the very heart of the wilderness in a southwest-northeast direction along the valleys of the Moose, Raquette and Saranac Rivers. Benedict proposed to develop a transportation network that would take advantage of the potential navigability of this natural 'slackwater" corridor. A group of investors from the iron district of Lake Champlain commissioned Benedict to conduct a detailed survey of the route in 1845. The following year, he submitted to the State Senate an untitled map of the survey, which outlined his scheme of a series of steamboat lines connected by short railroad portages running from Boonville in Oneida County to Port Kent in Essex County [#15]. A number of economic factors, both internal and external, prevented Benedict's plans from ever becoming a reality.

Another of Benedict's schemes met the same fate. In 1846, he proposed a canal from Long Lake to Round Lake to divert the headwaters of the Raquette River so that they would empty into the Hudson River. A purpose of "Benedict's Canal" was to facilitate the driving of timber down the Hudson. Benedict's proposal resurfaced in 1874, when the state canal commissioners sought to promote navigation on the Hudson River and the Champlain Canal. Engineer William B. Cooper conducted a trigonometric survey of the headwaters of the Raquette and Hudson Rivers from which draftsman C. D. Burrus made a *Map of Long, Forked and Little Forked Lakes. . .Map of Long Lake and Tributaries* [#16]. Lumbermen and manufacturers on the lower Raquette River were naturally opposed to the threat of this proposed diversion of their water, and the canal, though started, was never built.

Other ambitious proposals followed. Notable among the maps

Section of *Map of the Rail-Roads of the State of New York*, Sylvanus H. Sweet, 1864. This map is part of a series of state railroad maps which incidentally show- the interior Adirondacks as an area lacking in effective rail transportation throughout much of the nineteenth century. While the region's periphery is filled with railroad routes, only two proposed (but never completed) lines actually enter the central wilderness.

N
R. WAY
RIVER St LAWRENCE
St Regis
R. CHATEAUGAY
Chateaugay
Sharrington
Johnsons
Hemingford
Moores
Centerville
Black R.
Summit
Brandy Brook
Chazy
Ellenburgh
W. Chazy
Beekman
Burke
R. R.
Bangor
Brush's
Moira
Lawrence
Stockholm
Knapps
Potsdam Junction
Racket River
Madrid
Grass River
LISBON
NORTHERN
MALONE
NORTHERN R. R.
ENSBURGH
Deer R.
Salmon R.
CLINTON
PLATTSBURG
MOUNTS
Potsdam Village
Rensselaer Falls
CANTON
Oswegatchie R.
Hermon
De Kalb Junction
Richville
Governeur
LAWRENCE
St Regis R.
SARANAC RIVER
AU SABLE R.
LAKE CHAMPLAIN
PLACID L.
RACKET R.
SARANAC L.
Grass R.
Oswegatchie R.
CRANBERRY L.
DEAD L.
Mt SEWARD
Mt MARCY
ELIZABETHTOWN
ADIRONDAC IRON WORKS
LONG LAKE
L. DELIA
CROWN Pt
Hudson R.
Boreas R.
PARADOX L.
Natural Br.
Indian R.
BEAVER R.
RACKET L.
L. EMMONS
Schroon
4 Corners
TICONDEROGA
WIS
ADIRONDAC
FULTON CHAIN Ls
SCHROONS L.
Otter Cr.
RIVER
ESTATE
INDIAN L.
Turin
LYONS FALLS
MOOSE RIVER
Lyonsdale
Leyden Hill
Johnsburgh
Chester
BEAVER L.
INDIAN L.
WOODHULL RESR
CHUB L.R.
N.B. RESR
Talcottville
Hawkinsville
Woodhull
Williamsville
S.B. RESR
TRANSPt L.
Jesups R.
LAKE GEORGE
WHITE HALL
ROUND L.
LAKE PLEAS
L. PLASt
Warrens B.
CALDWELL
ONVILLE
CANAL
BLACK RIVER
Alder Cr.
PISECO L.
MOORE HOUSE
PROPOSED R.R.
Sacondaga R.
RAIL
ROAD
AND
WHITE HALL
Remsen
Prospect
Potsville
GLENS FALLS
MOREAU
Ft EDWARD
TRENTON
Trenton Falls
Holland P.
UTICA
Stittsville
ROME
Oriskany
Marcy
FISH L.
CAROGA L.
Ft MILLER
Batten Kill
Whitesboro
MOHAWK
Frankfort
Ilion
Herkimer
Little Falls
UTICA
FULTON
SARATOGA
SARATOGA SPGs
CANAL
CHAMPLAIN CANAL
St Johnsville
JOHNSTOWN
Tribes Hill
BALLSTON S.
Sauquoit
Clayville
LITTLE FALLS
RIVER
ERIE CANAL
Fort Plain
Palatine Bridge
Sprakers
Yosts
N. Y. CENTRAL
AMSTERDAM
Cranesville
Hoffmans
Johnsville
Hoosic R.
Columbia
Sangerfield
Crains Cor.
Starkville
Fulton
Ft Hunter
Jackson
MONTGOMERY
Sprakers Bas
MECHANICVILLE
TROY AND BOSTON
Winfield
Unadilla Forks
CANADERAGA L.
Canajoharie
Ames
Eatons Cor.
SCHENECTADY
ESPERANCE
Hyde
Exeter
OTSEGO L.
Burlington Flats
Oaksville
Sharon
Cobles Kill
SCHENECTADY
COHOES
LANSING BURGH CEMETERY
TROY
ARSENAL
N
W. Edmonson
COOPERSTOWN
Gardners V.
Barnerv
CEN. BR.
WILLIAMS HOLLOW
N. Y. CEN. R. R.
SARATOGA

that ensued was one by Abraham Franklin Edwards, the civil engineer for the Sacketts Harbor and Saratoga Railroad Company which had been incorporated in 1848 to market the iron and timber products of the interior Adirondacks. Edwards' 1853 map showed a proposed route following the valleys of the Grass, the Raquette, and the Hudson Rivers, from Saratoga to Sacketts Harbor. Twice reorganized and named the Adirondac Estate and Railroad Company in 1860, the company was controlled by the eminent railwayman, Thomas Clark Durant. Under Durant's aegis, the company built its road in several stages, its northern terminus reaching North Creek in 1871.

During this time, from the 1850s to the 1880s, the New York Legislature sponsored maps which showed the development of railroad and canal transportation in the state. The engineers and surveyors who compiled these maps revealed the Adirondack region to be a backwater of progress. Railroad lines circumventing the area were shown, but those for the interior never came to fruition. In contrast, most of the rest of New York State was animated by internal improvements, a network of operational railroads and canals. Engineer Sylvanus H. Sweet, one of the most prolific of state cartographers, represented this dichotomy of development on numerous maps, including his 1864 *Map of the Rail-Roads of the State of New York* [#17].

Settlers and Businessmen, 1850-1900

For a time, David Burr's county maps remained the most detailed cartographic account of settlement activity in the North Country. In 1853, Burr's work was joined by a *Map of the County of St. Lawrence*, compiled by the county historian and conservationist Franklin Benjamin Hough for his *History of St. Lawrence and Franklin Counties. . . .* Lot divisions, roads, railroads, villages, mines, mills, schools and churches were all important features on Hough's map.

The unprecedented detail and concentration of Burr's and Hough's maps set the stage for county wall maps of the late 1850s, the first to indicate the location and ownership of individual buildings. This type of map, called a cadastral map, appeared throughout the northeastern United States in response to a demand among businessmen, schools, and county officials for detailed plans of individual towns and counties. Advances in lithographic printing techniques made the new maps relatively inexpensive to manufacture.

Robert Pearsall Smith of Syracuse provided the impetus for the cadastral mapping boom in New York State. Negotiating contracts with local surveyors under the direction of superintendent John Homer French, Smith compiled survey information which various printers and publishers made into county wall maps. Copyrighted under Smith's name, these maps were cartographic equivalents to gazetteers. Each featured a large cadastral plan of a county, as well as insets of individual villages at larger scales. Some contained business and residence directories for major communities, illustrations of important buildings and natural features, and statistics from the latest state census (1855).

County wall maps of the 1850s provided valuable information about patterns of settlement, property ownership, physical configurations of villages, and economic and social activities of communities. County maps for the greater Adirondack region were: Washington (1853), Clinton (1856), Saratoga (1856), Fulton (1856), Lewis (1857), Herkimer (1857), Franklin (1858), Essex (1858), Warren (1858) and St. Lawrence (1858). Hamilton County lacked the population and businesses needed to make a cadastral map a profitable venture. The same held true for the northern part of Herkimer County, which was relegated to an insert on the 1857 wall map for the county. The best that mapmakers offered these adjoining areas was an 1851 *Map of Hamilton County and the North Part of Herkimer*, ordered by the board of supervisors of Hamilton County and compiled from original surveys made by William D. Jones. Although reasonably accurate, the Jones production lacked information of the kind shown on cadastral maps.

Insets in county wall maps displayed the spatial character of a number of Adirondack communities. Many settlements had developed along a river or road, others had formed at important crossroads, and still others had nucleated around a core of business and industry. Many Adirondack communities were dependent directly or indirectly on mining or logging. Commerce and agriculture figured in local economies as well. Cartographers also illustrated buildings and businesses which constituted the infrastructure of each town of size; these included schools, churches, attorneys, physicians, hotels, merchants, publishers, factories, farms, mechanics, banks and civic buildings. Tanneries, forges, furnaces, grist mills, saw mills and blacksmith shops also were featured.

The production of cadastral maps, interrupted by the Civil War, resumed after the conflict in the new format of county atlases of the late 1860s and 1870s. These new atlases contained a county map, village plans, and all the gazeteer-type accoutrements of the earlier wall maps. The compilers of the county atlases updated all statistical information and business directories, and modified the maps to allow for new information gleaned from original surveys by Silas N. Beers, Daniel G. Beers, Frederick W. Beers and other associates of the family business.

The Beers venture of New York City proved to be the most prolific of its kind, generating atlases of most Adirondack counties — St. Lawrence (1865), Saratoga (1866), Clinton (1869) **[#30]**, Lewis (1875), Franklin (1876), Washington (1876) and Warren (1876) **[#31]**. Joining the Beers family in the publication of county atlases of

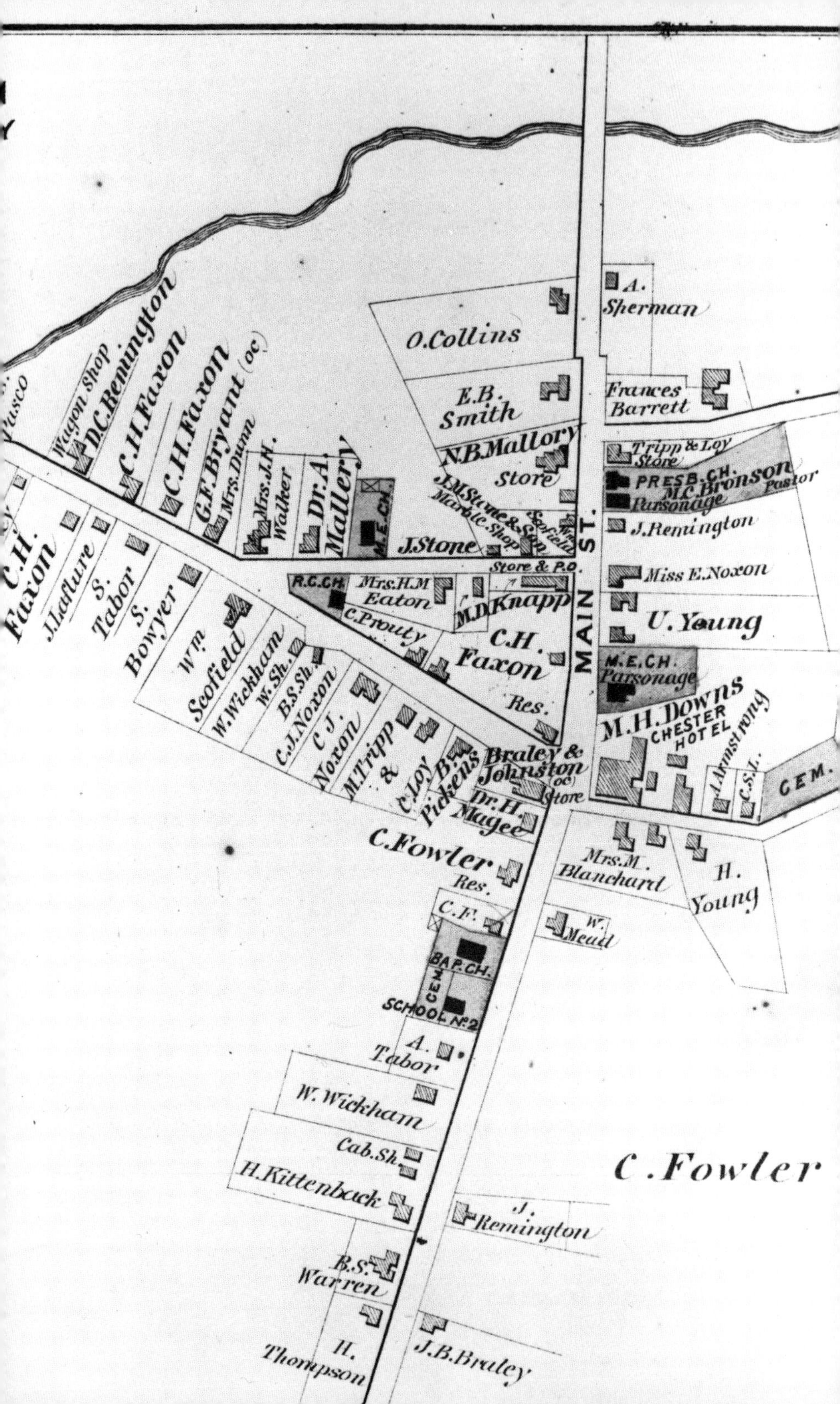

A. Sherman
O. Collins
E.B. Smith
Frances Barrett
N.B. Mallory Store
Tripp & Loy Store
PRES.B.CH.
M.C. Bronson Pastor
Parsonage
J.M. Stone & Son Marble Shop
Wm. Scofield
J. Remington
J. Stone
Store & P.O.
Miss E. Noxon
Wagon Shop
D.C. Remington
C.H. Faxon
C.H. Faxon
G.F. Bryant (oc)
Mrs. Dunn
Mrs. J.H. Walker
Dr. A. Mallery
M.E.CH.
Pasco
C.H. Faxon
J. Laflure
S. Tabor
S. Bowyer
R.C.CH.
Mrs. H.M. Eaton
M.D. Knapp
C. Prouty
U. Young
MAIN ST.
C.H. Faxon
Res.
M.E.CH. Parsonage
Wm. Scofield
W. Wickham
W. Sh.
B.S. Sh.
C.J. Noxon
C.J. Noxon
M. Tripp & C. Loy
B. Pickens
Braley & Johnston (oc) Store
Dr. H. Magee
M.H. Downs
CHESTER HOTEL
A. Armstrong
C.S.L.
CEM.
C. Fowler
Res.
Mrs. M. Blanchard
H. Young
C.F.
W. Mead
BAP.CH.
CEM.
SCHOOL No. 2
A. Tabor
W. Wickham
Cab. Sh.
H. Kittenback
C. Fowler
J. Remington
B.S. Warren
H. Thompson
J.B. Braley

northern New York were J. Jay Stranahan & Beach Nichols of New York City, who made the Herkimer County and Fulton/Montgomery Counties atlases in 1868, and Ormando W. Gray of Philadelphia, publisher of the Essex County atlas of 1876. Cadastral mappers once again bypassed the sparsely populated county of Hamilton.

The next cartographic format of Adirondack settlements was the perspective map, or birds-eye view. The lithographed panoramic map was universally popular among Victorians, perhaps because it depicted familiar cities and towns from a novel perspective. Artists were responsible for this type of map, drawing towns as if viewed from the air at an angle. Birds-eye views differed from earlier cadastral maps in that they were not planimetric drawings, but rather perspective illustrations without scale.

Lucien Rinaldo Burleigh of Troy drew panoramic maps of the larger and more affluent communities of the Adirondack periphery. His subjects included Ticonderoga (1884), Glens Falls (1884), Potsdam (1885), Canton (1885), Gouverneur (1885), Malone (1886), Port Henry (1889), Warrensburg (1891), and Plattsburgh (1899). In each of his birds-eye views, Burleigh sketched in the terrain of the village and laid on streets, buildings and parks with considerable fidelity to particular features. Each view carried a numbered key to a town's proudest structures and businesses. In so doing, the perspective map both celebrated and encouraged civic development as earlier cadastral maps and atlases had done. The artistry of Burleigh's portrayals appealed to town residents, who hung them in their homes and offices.

The mapping of Adirondack settlement and industry continued into the early twentieth century. Since the human landscape was always changing, cartographers continually updated and revised state and county wall maps and atlases. Theirs was a response to an ongoing demand for current information about communities by local businesses, real estate developers, insurance companies, public works projects, and the postal service. A map made for the latter purpose, the *Post Route Map of the State of New York. . .* of 1914, illustrated the spatial pattern and seasonality of mail routes. It also confirms a discernible pattern of Adirondack settlement – a series of "arms" extending inward from the periphery along major

---

"Chestertown," from *County Atlas of Warren, N.Y.*, Frederick W. Beers, 1876. Large-scale county atlas maps like this one illustrate in great detail the physical layout and the industrial, commercial and social activities of individual Adirondack communities. This is a "cadastral map" – it labels each building with the name of its owner.

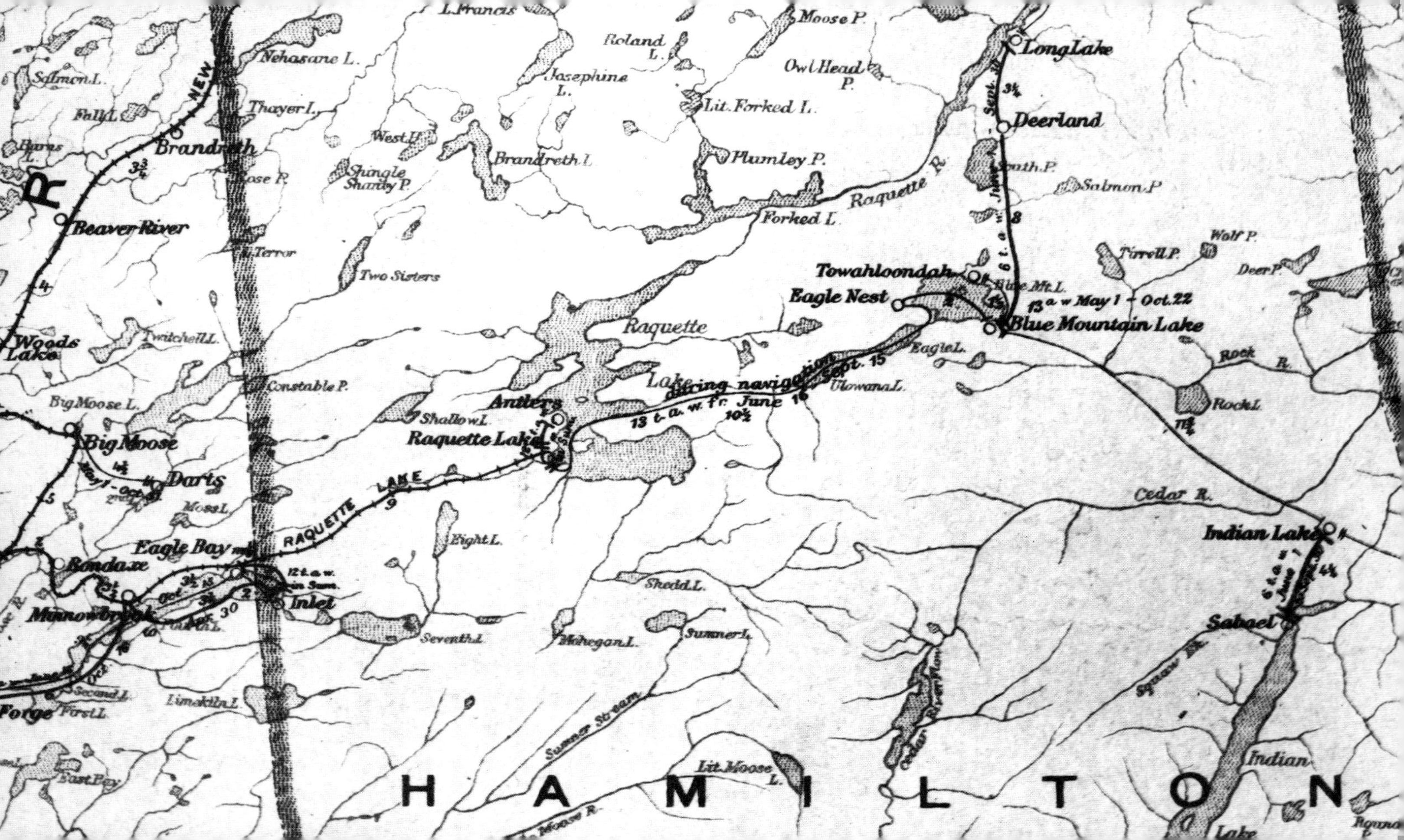
HAMILTON
Long Lake
Deerland
Towahloondah
Eagle Nest
Blue Mountain Lake
13 a w May 1 - Oct. 22
Raquette Lake
Antlers
13 t. a. w. fr. June 16
during navigation
Sept. 15
Raquette
Lake
RAQUETTE LAKE
Indian Lake
Sabael
Eagle Bay
Inlet
Minnowbrook
Rondaxe
Forge
Big Moose
Darts
Beaver River
Brandreth
Woods Lake
NEW
Nehasane L.
Thayer L.
Salmon L.
Fall L.
Two Sisters
Twitchell L.
Constable P.
Big Moose L.
Moss L.
Shallow L.
Eight L.
Seventh L.
Limekiln L.
Second L.
First L.
East Bay
Shedd L.
Mohegan L.
Sumner L.
Sumner Stream
Lit. Moose L.
Cedar R.
Cedar River Flow
Squaw Bk.
Indian
Lake
Rock R.
Rock L.
Eagle L.
Utowana L.
Blue Mt. L.
Tirrell P.
Wolf P.
Deer P.
Salmon P.
Raquette R.
Forked L.
Plumley P.
Lit. Forked L.
Owl Head P.
Moose P.
Roland L.
Josephine L.
L. Francis
Brandreth L.
West H.
Shingle Shanty P.
Terror

transportation routes [#32].

By the time these later maps arrived on the scene, in the years 1890-1920, the era of the "Advancing Frontier" was drawing to a close. Other maps, involving a totally new perception of the Adirondack region, had established themselves and had begun to overshadow the cartographic themes of economic development and commercial exploitation.

Section of *Post Route Map of the State of New York. . .*, United States Post Office Department, 1914. This map shows how mail travelled to and from the Adirondacks by road, railroad and steamboat. Special notes on the map indicate the frequency and seasonality of mail delivery to villages and summer tourist spots.

The Wilderness Resort

Tourists and Sportsmen, 1860-1890

Beginning in the 1830s, writers and travelers expressed romantic perceptions of the Adirondacks in the popular press. Charles Fenno Hoffman, Joel Tyler Headley, Samuel H. Hammond, Charles Lanman, and the members of the Philosophers Camp at Follansbee Pond waxed eloquent about the peaceful seclusion, the healthful climate, the picturesque beauty, and the invigorating field sports of the North Country. Other writers followed, some of them choosing to embellish their travel accounts with maps.

The first tourist maps of the Adirondacks were really illustrations for the writings of the backwoods humorist Thomas Bangs Thorpe, in *Harper's New Monthly Magazine* for 1859; William Watson Ely, a vacationing physician, in *Moore's Rural New Yorker*, 1860; and the writer Alfred Billings Street, in *Woods and Waters*, 1860. None of these men were professional mapmakers, but their love of the Adirondack wilderness and their quest for intellectual pursuits qualified them as enthusiastic amateurs. The popular recreational route described by these men and by Edwin Atkins Merritt was the same route of slackwater navigation between Boonville and Port Kent surveyed fifteen years earlier for commercial and industrial purposes by Farrand Benedict.

Merritt, a civil engineer who had spent the previous decade conducting surveys for internal improvements in St. Lawrence County, responded to a demand among sportsmen and vacationers who needed an accurate map of the Raquette River Valley. His *Map of the Racket River between Stark's Falls and Tupper's Lake* and *Map of the Head Waters of the Racket River*, published in 1860, illustrated boat portages, camps, a hiking trail, and areas with good trout fishing and deer hunting.

Merritt also made an early regional map of the Adirondacks with the literary traveler, Homer de Lois Sweet, in 1867. *The Great Wilderness of Northern New York and a Sketch of the Border Settlements* noted the location of guides' quarters, hotels and supplies for the recreational tourist. This comprehensive map was one of the first to feature the entire Adirondack region alone as its sole subject. It appeared in at least one revised edition, but was overshadowed by

St. Regis River
Paul Smith
Follensbee P.
Spitfire Pond
Spectacle Ponds
Bay P.
St. Regis Lake
Fish P.
Cat P.
Ochre P.
St. R. P.
BRANDON
Big Clear P.
Floodwood P.
Indian L.
Jordan River
Deer P.
DICKINSON
Willis P.
Rollins P.
B. Square P.
El. 1527 ft.
Moosehead
Stillwater
BLUE MT
L. Long Pd.
L. Rock P.
Big Wolf Pond
Lead P.
Saginaw Bay
HARRIETSTOWN
Hungry Bay
Joe P.
Hedgehog Rapids
Pitchfork P.
Massawepie Lake
Little Wolf Pond
Bartlett's
Half way Brook
Cory
Long P.
Raquette Pond
Stoney Cr. Pds
Ampersand P.
Gull P.
ARAB MT
O'Loughlin
Stetson
Raquette River
Calkins
Stoney Cr.
Simons P.
L. Simons P.
Follensbee P.
Johnson's
Raquette Falls
El. 1545 ft.
TUPPER LAKE
Bridge Brook P.
Horseshoe Pond
Graves
Reeds Bay
Herricks Bay
Moose Cr.
Bog River
Falls
Jenkins Pd.
Duck P.
Anthony Pond
Cold River
SANDANONA MT
Round Pond
Handsome P.
50
Round P.
Moose P.
L. Harkness
LITTLE TUPPER L.
23
Mohegan P.
22
Stoney P.
Grampus Lake
LONG L.
El. 1575 ft.
Slim P.
Long P.
Mud P.
Big Brk
Kellogg
LAKE
Pickwacket Pond
Catlin Lake
NEW
L. Delia
Rock Pond
Clear Pond
C. B. Kellogg
ROAD
36
Bottle Pond
Smith
Mitchell
Sabbattis
20
Sutton P.
21
Rich P.
L. Harris
Carey P.
Moose P.
OWLS HEAD
El. 2700 ft.
Little Forked Lake
Palmers
MT GOODWIN
Carys
35
Plumley P.
Tirrell P.
Forked Lake
El. 1704 ft.
South P.
Hudson
Hotel
18
Minnie Pond
BLUE MOUNTN
Sargents Ponds
Chain Lake
Beaver Cr.
BLUE MOUNTAIN L.
El. 1791 ft.
Ned Buntline
Eagle L.
Utowana L.
34
Stephen P.
Rock River
Loon Cr.
17

the enormous popularity of another regional map published the same year by William Ely.

Ely's *Map of the New York Wilderness* was aimed at lovers of sport and romantic scenery [#38]. Although the map included the region's periphery, Ely concentrated on the lakes district centering on Raquette and Long Lakes, and the High Peaks region surrounding Keene Valley, both areas of interest to tourists. Publishers George Woolworth Colton and Charles B. Colton of New York City marketed Ely's map in a pocket format, folded between two cardboard covers for the convenience of the traveler.

Ely's map, though not as original or complete as the Sweet and Merritt production, became the standard map for most Adirondack tourists during the next dozen years. A portion of the 1868 edition was reproduced as *Map of the Saranac Lakes* for D. L. Fouquet & Son in 1869. The minister William H. H. "Adirondack" Murray used a version of the 1869 edition for his book, *Adventures in the Wilderness*. This publication often is credited as sparking the tremendous growth in regional tourism which began after the Civil War. Among guidebooks to carry later editions of the Ely map were: Winslow Cossoul Watson's *A Descriptive and Historical Guide to the Valley of Lake Champlain and the Adirondacks* (1871); Edwin R. Wallace's *Descriptive Guide to the Adirondacks* (from 1872 through 1897); and Seneca Ray Stoddard's *The Adirondacks, Illustrated* (from 1874 through 1879).

A deluge of Adirondack tourist maps followed Ely's, and these featured important information on routes and fares of stages, railroads and steamboat lines, and the location of hotels, camps and guides' quarters. Edwin R. Wallace and Seneca Ray Stoddard became the most prolific producers of travelers' maps in the last decades of the nineteenth century. By 1880, when Stoddard began making his own *Map of the Adirondack Wilderness*, the North Country had become a fashionable resort, offering a positive "wilderness experience" for many well-to-do vacationers from New York City and other urban areas of the East [#40].

A lakeside resort area stretching on an axis from Old Forge to Upper Chateaugay Lake had established itself in the Adirondacks by 1880. Hotels great and small were built on lake shores, and stagecoaches and steamboats connected one resort community to its neighbors. Regional tourist maps illustrated this central lakes

---

Section of *Map of the New York Wilderness*, William Watson Ely, 1867. This is one of the earliest regional maps made for tourists to the Adirondacks. The map notes the locations of hotels, guides' quarters and travel routes in the region's central lake district.

district, though several, more localized maps appeared as well. The latter included *The New York Wilderness. Hamilton County and Adjoining Territory*, compiled by Benjamin Clapp Butler in 1879, and three maps published during the 1880s in Frank H. Taylor's guidebook, *Birch Bark from the Adirondacks*.

Railroad executives recognized the potential of tourist travel by shifting the emphasis of their maps from freight to passenger transportation. Benjamin Butler, who was one of the most active compilers of railroad tourist maps, produced a series of pamphlets in the 1870s and 1880s that illustrated what he called the "Great Pleasure Route" of the Delaware and Hudson Railroad along the eastern fringe of the Adirondacks [#42]. Cartographers were commissioned to produce passenger maps for other upstate railroads – the Adirondack Railroad, the Ogdensburgh and Lake Champlain Railroad, the Utica and Black River Railroad, the Rome, Watertown and Ogdensburgh Railroad, and the Chateaugay Railroad [#44]. A number of these promotional railroad maps found their way into popular guidebooks before the turn of the century.

The central Adirondacks had acquired the reputation of being a paradise for rugged sports; by contrast, the Champlain Valley had long attracted tourists interested in less stressful vacations that centered around historic sites, scenic wonders, steamboat cruises and other genteel activities considered suitable to women and children. Maps of this area appeared in travel guides published by Taintor Brothers and Company, Asher and Adams, and C. A. Faxon. Lake George, the valley's oldest and most successful resort, was the subject of many tourist maps after 1870. The Beers family published a *Map of Lake George & Vicinity* which was derived from their Warren County and Washington County atlases of 1876. Seneca Ray Stoddard, author of the guidebook, *Lake George Illustrated*, issued two series of Lake George tourist maps throughout the 1880s and 1890s [#41]. During the same period, Charles H. Possons of Glens Falls published fold-out guidebook maps of Lake George and Lake Champlain.

Adirondack tourist maps touted a region offering pure air and

---

Section of *Map of the Delaware and Hudson Canal Co.'s Railroads and Connections*, Benjamin Clapp Butler, 1879. This map, published by a passenger railroad company, reflects the growing Adirondack tourist industry that began to develop shortly after the Civil War. The map shows train routes (solid black lines), railroad stops (white dots), and connecting stage and steamboat lines (dashes).

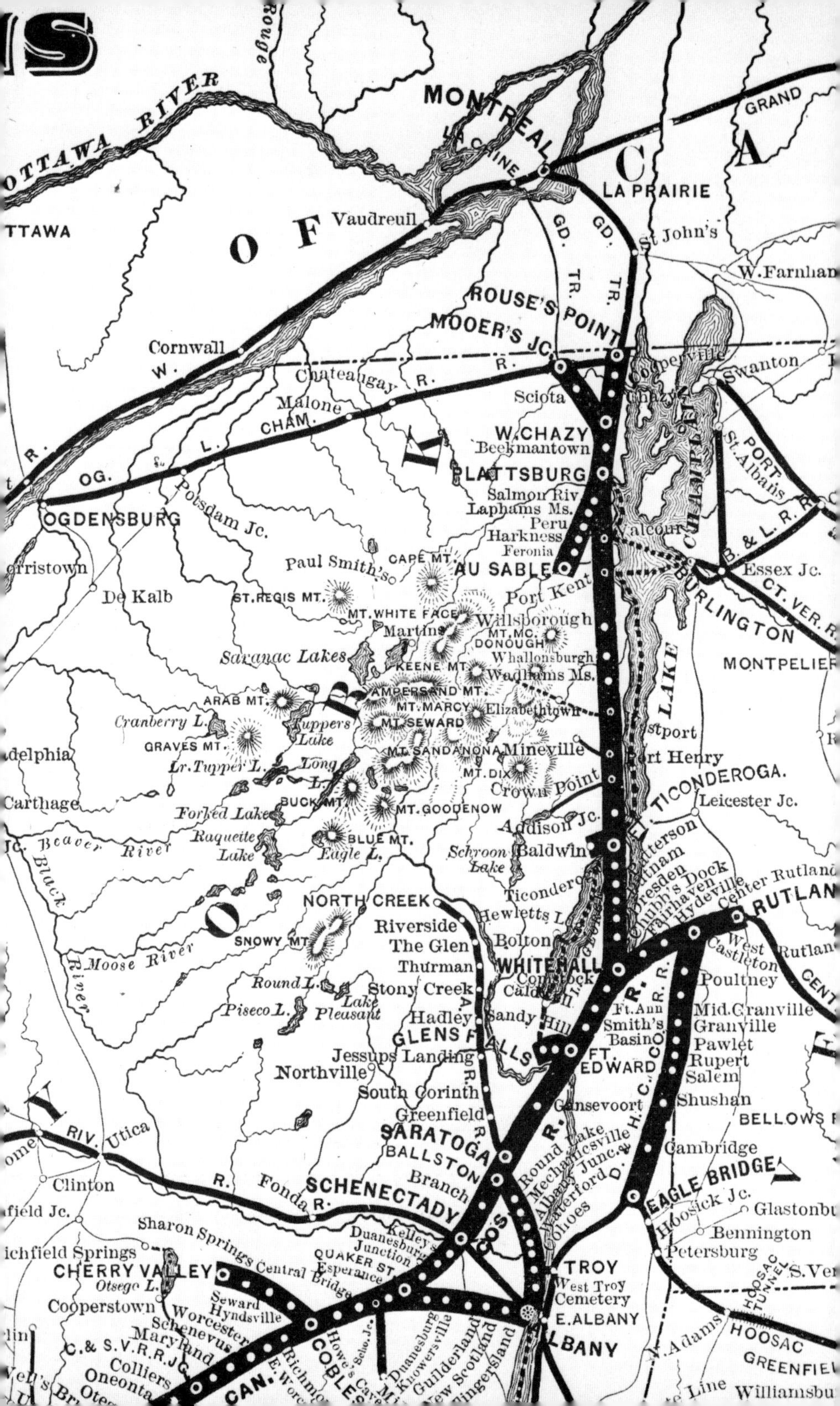
S
Rouge
OTTAWA RIVER
MONTREAL
LACHINE
GRAND
C
A
LA PRAIRIE
TTAWA
Vaudreuil
O F
GD. TR.
GD. TR.
St John's
W. Farnham
ROUSE'S POINT
MOOER'S JC.
Cornwall
W.
R.
Chateaugay
R.
R.
Cooperville
Swanton
Sciota
Chazy
Malone
CHAM.
L.
W. CHAZY
Beekmantown
PORT
St. Albans
R.
OG.
PLATTSBURG
Salmon Riv.
Laphams Ms.
OGDENSBURG
Potsdam Jc.
Peru
Harkness
Feronia
Walcott
CHAMPLAIN
B. & L. R. R.
Paul Smith's
CAPE MT.
AU SABLE
Essex Jc.
orristown
De Kalb
ST. REGIS MT.
Port Kent
BURLINGTON
CT. VER. R.
MT. WHITE FACE
Willsborough
Martins
MT. MC. DONOUGH
Saranac Lakes
Whallonsburgh
MONTPELIER
KEENE MT.
Wadhams Ms.
LAKE
AMPERSAND MT.
ARAB MT.
MT. MARCY
Elizabethtown
Cranberry L.
Tuppers Lake
MT. SEWARD
GRAVES MT.
Westport
delphia
Lr. Tupper L.
MT. SANDANONA
Mineville
Port Henry
Long L.
MT. DIX
TICONDEROGA.
Crown Point
Carthage
Leicester Jc.
BUCK MT.
MT. GOODENOW
Forked Lake
Addison Jc.
Raquette Lake
BLUE MT.
Patterson
Beaver River
Eagle L.
Baldwin
Putnam
Black
Schroon Lake
Dresden
Chubb's Dock
Ticonderoga
Fairhaven
Hydeville
Center Rutland
NORTH CREEK
Hewletts L.
RUTLAND
Riverside
West Castleton
O
SNOWY MT.
The Glen
Bolton
Rutland
Moose River
Thurman
WHITEHALL
Comstock
Poultney
Round L.
Stony Creek
Caldwell
R.
River
Lake Pleasant
Piseco L.
Hadley
Sandy Hill
Mid. Granville
Granville
Ft. Ann
Smith's Basin
GLENS FALLS
Pawlet
Jessups Landing
FT. EDWARD
Rupert
Northville
Salem
South Corinth
Shushan
Gansevoort
Greenfield
BELLOWS F
Y
RIV.
Utica
SARATOGA
R.
BALLSTON
Round Lake
Mechanicsville
Albany Junc.
Cambridge
Clinton
R.
Fonda
Branch
SCHENECTADY
D. & H. C. CO. R. R.
EAGLE BRIDGE
field Jc.
Waterford
Cohoes
Hoosick Jc.
Glastonbu
Sharon Springs
Duanesburg
Kelley's Junction
Bennington
ichfield Springs
QUAKER ST.
Esperance
Petersburg
CHERRY VALLEY
Central Bridge
TROY
HOOSAC TUNNEL
S. Ver
Otsego L.
West Troy
Cemetery
Seward
Hyndsville
Cooperstown
Worcester
E. ALBANY
Schenevus
Maryland
Howe's Cave
Duanesburg
Knowersville
Guilderland
New Scotland
ALBANY
N. Adams
HOOSAC
C. & S. V. R. R. JC.
Colliers
Oneonta
CAN.
COBLES
Richmond
GREENFIE
Williamsbu

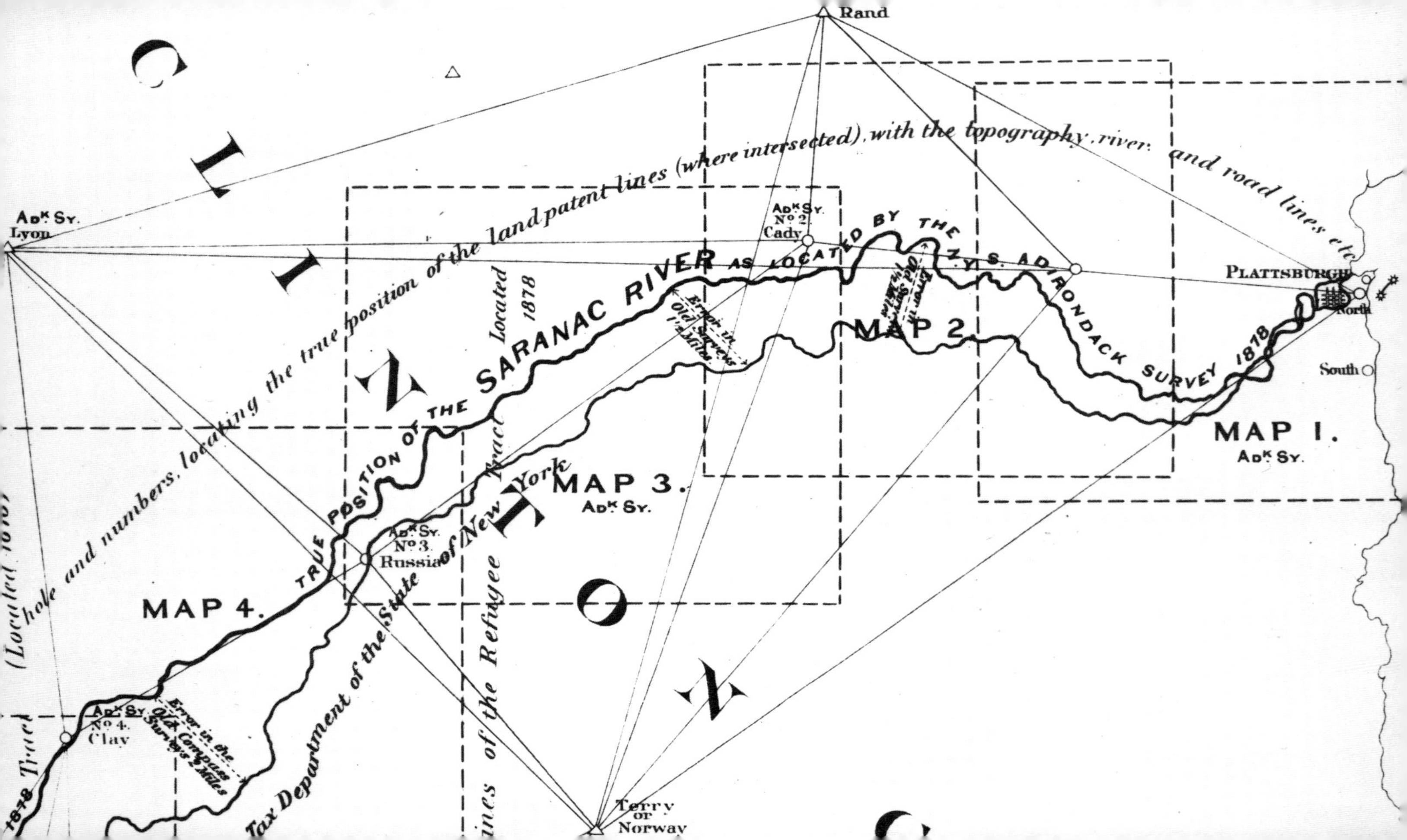

CLINTON C
Rand
Adk. Sy. Lyon
Adk. Sy. No. 2. Cady
Adk. Sy. No. 3. Russia
Adk. Sy. No. 4. Clay
Terry or Norway
Plattsburgh
North
South
MAP 1. Adk. Sy.
MAP 2
MAP 3. Adk. Sy.
MAP 4.
THE SARANAC RIVER AS LOCATED BY THE N.Y. S. ADIRONDACK SURVEY 1878
TRUE POSITION OF THE SARANAC RIVER
Located 1878
Error in Old Surveys 1½ Miles
Error in Old Surveys 1½ Miles
Error in the Old Compass Surveys 2 Miles
(Located 1878)
hole and numbers, locating the true position of the land patent lines (where intersected), with the topography, river, and road lines etc.
Tax Department of the State of New York
ines of the Refugee Tract
1878 Tract

water, healthful exercise, beautiful scenery, and restful solitude. Their appeal was to a growing constituency of urbanites who were far less concerned with marketable minerals and timber than with travel arrangements and accomodations for themselves and their families. To these maps were added by 1890 a newer type, the map of conservationists urging the protection of New York's northern forest for utilitarian rather than recreational purposes.

Scientists, Engineers and Conservationists, 1872-1918

By almost any measure, the Adirondacks of the post-Civil War period was underdeveloped; its settlement, industry and transportation lagged far behind the rest of New York State. The era of the "Advancing Frontier," however, had fostered a materialistic, man-over-nature value system which tolerated and abetted wholesale, destructive manipulation of the Adirondack environment. The laissez-faire legislation of an industrial America endangered the valuable ecosystem of the northern wilderness, depleting its soil and its timber and water reserves.

Into these circumstances arrived Verplanck Colvin, a land surveyor who, like the scientists, engineers and literary tourists who preceded him, appreciated the Adirondack region for its own sake. After undertaking several investigations at his own initiative and expense, Colvin convinced the New York Legislature to support a systematic topographical survey of the Adirondack wilderness in 1872. As superintendent of the Adirondack Topographical Survey, Colvin made a number of noteworthy advances in the mapping of the region's landform. The many field maps which Colvin included in his reports to the legislature in the 1870s demonstrated the strides he was making in charting the topography of the North Country.

Armed with theodolite, transits, leveling rods, and the principles of trigonometry, Colvin charted elevations in the Adirondacks with a precision unattainable with earlier barometric readings. His familiarity with the lay of the land enabled him to use contour lines to depict land relief. The contours and vertical hatchings of maps such as the

---

Section of *Sketch Showing the Saranac River Survey. . .*, Verplanck Colvin, 1878. This map reflects the accurate survey work accomplished by Verplanck Colvin. It shows the difference between the course of the Saranac River as charted by previous surveyors and "the true position" of the river as located by Colvin's Adirondack Topographical Survey. The lines drawn between mountain-tops illustrate Colvin's "triangulation" method of surveying.

Secondary Reconnaisance Sketch of Mount Marcy. . . of 1873 provided the readers of Colvin's reports with a fresher understanding of Adirondack geography than had been attained previously [#36].

Colvin corrected errors in earlier official maps used by the tax department of New York State by shunning the use of magnetic compasses whose readings were affected by large deposits of iron-bearing ore. From the many mountaintop survey posts he established, Colvin mapped out a complex network of triangles which yielded the true locations of many of the region's natural features relative to one another. This method of triangulation produced maps such as *Sketch Showing the Saranac River Survey (Base Lines) 1878*, which illustrates the contrasting results of Colvin's readings and those of older, less accurate, compass-based surveys [#37]. For his triangulations, Colvin used a base line established by the United States Coast and Geodetic Survey in its concurrent study of Lake Champlain. The federal government's large-scale map of Lake Champlain in 1880 represented the highest degree of cartographic knowledge of the lake available to that time.

Colvin's achievements establish him as a pivotal figure in the history of Adirondack mapping. Mapmakers such as Seneca Ray Stoddard quickly adopted the revised information gleaned from Colvin's surveys of the 1870s. Verplanck Colvin's significance, however, transcended his work for the Adirondack Topographical Survey; he also contributed to rising conservationist sentiment which led to the formation of the New York State Forest Preserve in 1885 and the Adirondack Park in 1892.

Colvin and others feared that the destructive and wasteful practices of the logging industry would seriously deplete the region's valuable timber and its great watersheds. A new pulp and paper industry which encouraged clearcutting added to the apprehensions of these early conservationists. In 1883, New York State terminated the practice of selling its Adirondack properties to private buyers; too often these purchasers had exploited the land as best they could and then had allowed their tracts to revert back to the state for unpaid taxes. That year the legislature appointed Verplanck Colvin superintendent of the State Land Survey to determine the location and extent of state-owned lands in the Adirondacks and the Catskills.

For the next fourteen years, the administrative surveys carried out by Colvin and his assistants generated a number of maps showing in detail the accurate locations of boundaries of old townships, tracts and lots involving state lands throughout the northern wilderness. The project was not confined to isolated mountains and lakes of the region's interior, as the Adirondack Topographical Survey had been. Some of the maps from the State Land Survey, such as the 1897 *Map Showing the Location of the Highway Between Northville and*

Wells, covered more settled peripheral territories as well [#34]. The New York State Land Survey, the United States Coast and Geodetic Survey, and the United States Geological Survey, which began its work in the Adirondacks during the 1890s [#38], provided cartographers with a remarkably accurate and detailed base for their maps and atlases.

The newly-formed New York State Forest Preserve of 1885 introduced a new type of map designed to depict environmental conditions. The earliest of these, appearing in the first annual report of the New York State Forest Commission in 1885, was titled, *Map of the Adirondack Plateau Showing the Position & Condition of Existing Forests*. It divided the North Country into regions of virgin forest, merchantable timber, denuded territories and farmlands, all superimposed over a patchwork of state-owned land. The commission published another edition of this map in 1890, but with the significant addition of the famous "Blue Line," as indicated in the title, *Map of the Great Forest of Northern New York showing boundaries (in red) of the Forest Area and boundaries (in blue) of the Proposed Adirondack Park* [#55]. The Adirondack Park proposed in the map would become a reality in 1892, its purpose being to define the area within which the State would concentrate its land purchases for the purpose of conservation. From this time onward, the "Blue Line" which delineated the Park became a distinguishing feature in maps of the Adirondacks. Another step forward in the protection of the Adirondack wilderness was taken in 1894, when the state legislature passed a constitutional amendment to keep Forest Preserve lands in the Adirondack Park "forever wild."

In this atmosphere of environmental protection, the New York State Forest Commission and its successor after 1895, the Forest, Fish and Game Commission, produced regional maps stressing the importance of the Adirondacks. At times, the commissions relied upon outside sources for general reference maps bound into their annual reports: the tenth edition of Seneca Ray Stoddard's tourist-oriented *Map of the Adirondack Wilderness* appeared in the Forest Commission's report for 1888, and Julius Bien's *Map of the Tracts, Patents and Land Grants of Northern New York* was reproduced in the 1893 report.

State-produced resource management maps of the Adirondacks between 1884 and 1918 dealt with several conservationist themes, each involving the public control of natural resources implicit in America's Progressive Era. These themes were public land administration, scientific forestry, forest protection, and watershed regulation and development.

The most significant administrative maps of the period were large-scale wall charts, issued as four-section pocket maps and updated every few years by various state conservation agencies. (The latest

BLUE MOUNTAIN L. P.O.
CHAIN LAKES
Cedar R.
INDIAN LAKE
MINERVA
JACKSON
INDIAN LAKE P.O.
Bad Luck P.
WASHINGTON DAM
INDIAN LAKE
SCHROON RIV. P.O.
BLUE RIDGE
OLD FURNACE
HAMMONDVILLE
PARADOX LAKE
SCHROON LAKE P.O.
SCHROON LAKE
PHAROAH MT.
PHAROAH L.
Sucker Brook
MINERVA P.O.
OLMSTEADVILLE
JONES MT.
ADIRONDACK P.O.
PARK MT.
TRUMBULL MT.
NORTH RIVER P.O.
DAM
THIRTEENTH P.
BULLHEAD MT.
Botheration P.
NORTH CREEK
Hudson River
POTTERSVILLE P.O.
CHESTER
Valentine P.
BRANT LAKE
Puffer Pond
LOON LAKE
RIVERSIDE
WEAVERTOWN P.O.
JOHNSBURG
FRIENDS LAKE
HORICON
CHESTERTOWN P.O.
SABBATH DAY P.O.
SNOWY MT.
Overflow
Willow Br.
DUG MT.
Elm L.
LAKE PLEASANT
OREGON P.O.
CRANE MT.
THE GLEN P.O.
SPRUCE MT.
NORTH BOLTON
FEDERAL HILL
NORTH WEST BAY
MT. EREBUS
SHELVING ROCK
SLEEPING BEAUTY
THURMAN P.O.
BOLTON P.O.
Shires P.
Round P.
Indian P.
Mud P.
30
26
25
16
15
33
32
31
13
12
24
11
29
9
8
34
5
4
3
10

edition is dated 1983.) The first title in the series, published in 1893, was *Map of the Adirondack Forest and Adjoining Territory Compiled From the Official Maps and Field Notes on file in the State Departments at Albany, N.Y.*, but later editions, from 1938 onward, were called simply *Adirondack Map*. The series delineated the Adirondack Park and Forest Preserve lands in great detail and provided a valuable reference tool for a state government increasingly engaged in conservation efforts in the North Woods.

Conservationists advocated the replacement of wasteful clearcutting of Adirondack woodlands with scientific management which would yield sustained harvests of timber. The planning of careful and efficient logging practices required maps such as H. S. Meekham's 1901 *Map of Townships 5, 6, 40 and 41, Totten & Crossfield Purchase, Hamilton County, N.Y.*, which noted lumbering activity, merchantable woodlands, and land ownership [#57]. Scientific forest management resulted in maps which illustrated the distribution of forest types and conditions; one of these was *Map of Warren County Showing Forest Types*, made for the New York State Conservation Commission in 1911.

The prevention of forest fires was a priority among Adirondack conservationists at the turn of the century; a number of maps reflected the measures taken by the state to reduce destruction from fires caused by railroads and other users of the woodlands. Among these were the Conservation Commission's *Map of Adirondack Preserve Counties* of 1911 [#58], and *Fire Protection Map of the Adirondack Forest* of 1916. The latter showed in detail the impact of the conservationist ethic in northern New York, since it divided the forest cover into areas of green timber, burned timber, softwood and hardwood logging, waste and denuded lands, and open land. It also noted forest ranger camps, storage for fire fighting tools, fire observation stations, access roads, and a telephone communications network vital to the reporting of forest fires.

Concern for the watershed played a large role in the production of Adirondack maps by scientists and engineers during the late 1800s and early 1900s. The prevention of erosion and flooding, the maintenance of a sustained stream flow and water supply, and the devel-

---

Section of *Map of the Great Forest of Northern New York...*, A. Robeson, 1890. This conservationist's map was one of the first to outline the boundaries of the Adirondack Park, shown as a "blue line" (the dark line). Various shadings represent the conditions of the forest cover – virgin, harvested, and denuded – while the checkerboard pattern indicates the extent of state-owned land.

opment of sites for hydroelectric power occupied the attention of these mapmakers. Since each dam and reservoir project required a map, cartographers provided a detailed record of state regulatory efforts. This record included: *Map of Water Shed Supplying Reservoirs in Adirondack Forest*, made for the state engineer and surveyor in 1898; and a map of *Oswegatchie River – East Branch, Proposed Newton Falls Reservoir*, prepared for the New York State Conservation Commission in 1913 [#56 and #60].

Visitors and Seasonal Residents, 1890-1930

Between 1890 and 1920, sportsmen and nature lovers continued to rely upon the types of guidebook tourist maps which had first become popular after the Civil War. Seneca Ray Stoddard remained the single most prolific Adirondack mapmaker for urban travelers, and his regional and Lake George maps persisted until a few years before his death in 1917.

Railroad companies continued to publish promotional tourist maps. The Delaware and Hudson line was perhaps the most publicized tourist route to the Adirondacks, but stiff competition came from the railroad built by Dr. W. Seward Webb. Completed in 1892, Webb's Mohawk and Malone Railway was the first to cross the Adirondacks, penetrating the heart of the northern wilderness between Herkimer in the south and Malone in the north. Acquired by the New York Central in 1893, the railroad changed the face of tourism in the central Adirondacks by offering easy access to the region and delivering a fresh influx of tourists to the least developed areas of the wilderness. The New York Central Railroad churned out a number of tourist maps in brochures boasting about the recreational and scenic benefits of the "central lake region of the Adirondack Mountains."

Groups of individuals banded together in the late nineteenth century to purchase large tracts of land for social and recreational purposes. Exclusive, yet conservationist in their management, the preserves established by these clubs were featured prominently on regional maps of the 1890s. A number, including the Adirondack League Club, the Vilas Preserve, the Adirondack Preserve Association and the North Woods Club, had large-scale maps made to show their

---

Section of *Map of Arbutus Camps. . .*, Wesley Barnes, 1899. This manuscript map represents the cartographic needs of a growing number of wealthy seasonal residents in the Adirondacks toward the end of the nineteenth century. It shows the layout of a private woodland "camp" complex planned by William West Durant.

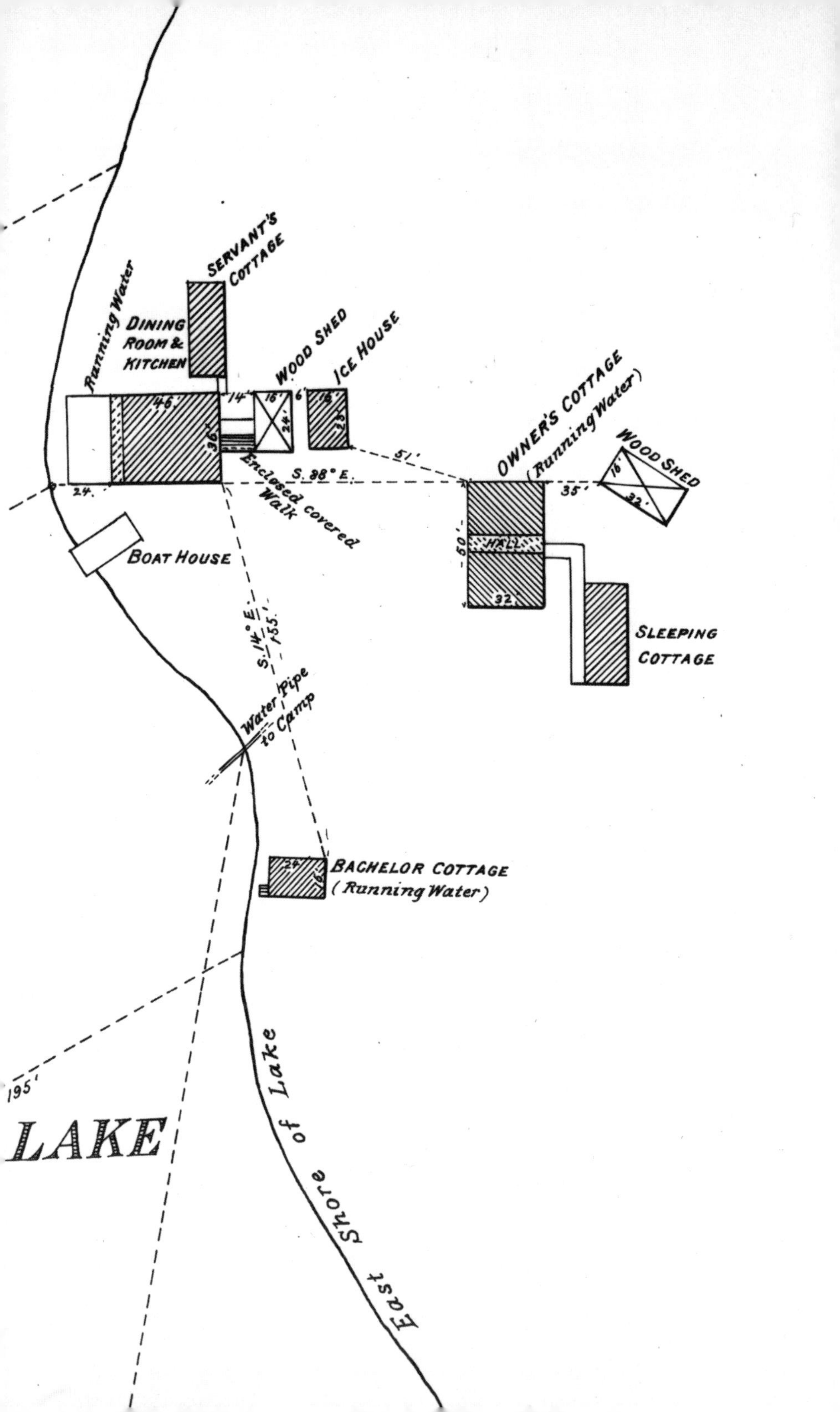

SERVANT'S COTTAGE
Running Water
DINING ROOM & KITCHEN
WOOD SHED
ICE HOUSE
46'
14'
16'
6'
16'
24'
23'
36'
24'
OWNER'S COTTAGE (Running Water)
WOOD SHED
16'
32'
51'
S. 88° E.
35'
Enclosed covered Walk
BOAT HOUSE
HALL
50'
32'
SLEEPING COTTAGE
S. 14° E.
155'
Water Pipe to Camp
BACHELOR COTTAGE (Running Water)
24'
16'
195'
LAKE
East Shore of Lake

GASOLINE & AUTO SUPPLIES
Opposite the Ondawa.
G. W. Taylor,
Schroon Lake.
AIDEN LAIR
AIDEN LAIR LODGE
M. F. Cronin, Prop. The starting point of Roosevelt's famous ride, September 14, 1904, with M. F. Cronin, driver.
ONDAWA HOUSE
Headquarters for Automobilists. Open the year round. New Garage. F. C.
SCHROON LAKE
LELAND HOUSE
ONDAWA HOUSE
TAYLOR HOUSE and COTTAGES.
C. A. Taylor, Jr. $4
Gasoline, Oils, Repairing. Free Garage.
THE LELAND HOUSE
Fire Proof Garage. Supplies. C. T. Leland, Mgr.
TAYLOR'S-ON-SCHROON
EAGLE LAKE
TICONDEROGA
Clay
ROGERS ROCK HOTEL
ROGERS ROCK
BALDWIN
ORWELL
LARRABEES
LAKE HOUSE
Larrabees Point
Frederick Ives,
$2.50 up.
Unsurpassed views of the Lake. Automobile accommodations. Boating and Woodland Walks. Season June 1 to September 15. Address Rogers Rock Hotel, Rogers Rock, N. Y.
ISLAND HARBOR
HAGUE
LAKE GEORGE STEAMBOAT
Carries Auto and Driver Between Bolton and Sabbath Day Point for $2.50 to $3.50
SILVER BAY
SABBATHDAY POINT
Big hill
Big hill
HULETT'S LANDING
ISLAND HARBOR.
House and Cottages
Hague-on-Lake George.
B. A. Clifton, Prop.
Rates $2.50 up.
TONGUE MOUNTAIN
BLACK MOUNTAIN
SAGAMORE-ON-LAKE-GEORGE.
Rates $5 up. Most ideal resort on the trip.
FAIR HAVEN
STEWART'S PEBLOE.
H. Ward Stewart, Jr.
$3 up. Garage.
Supplies.
STEWART'S PEBLOE
BRANT LAKE
MINERVA
NORTH RIVER
POTTERSVILLE
NORTH CREEK
ADIRONDACK HOUSE
LOON LAKE
RIVERSIDE
CHESTERTOWN
CHESTER HOUSE
WEVERTOWN
THE GLEN
CHESTERTOWN
7 MILES
7 M.
5 MILES
5 MILES FAIR
6 MILES FAIR
9.5 M.
6 M.
12 M.
5 M.
9 M.
ADIRONDACK HOUSE,
North Creek, N. Y.
Accommodations for Automobilists $2.50 up.
CHESTER HOUSE,

forest and lakeside properties, lodges and clubhouses [#46].

Maps at this time reflected the appeal which the Adirondacks held for well-to-do Americans at the turn of the century. William West Durant, for example, supervised the construction of large woodland estates which he sold to wealthy urbanites. Following the example of his father, Thomas Clark Durant, William improved and expanded his family's holdings in the Adirondacks, thereby becoming the region's first large-scale developer. He employed the surveyor Wesley Barnes, the draftsman M. van Mittendorfer, and others to map the tracts on which he hoped to build extensive retreats for the wealthy and smaller ones for middle class families of more modest means. Durant's schemes produced a voluminous output of manuscript maps, principally of areas near Raquette Lake, Blue Mountain Lake and Newcomb. Presenting information needed for planning, selling, insurance and other purposes, these maps inventoried the physical layout of each property. Subjects included the famous "great camps" of Pine Knot, Uncas, Sagamore and Kill Kare, as well as Durant's properties near Newcomb, which included Goodenow Mountain, Zack Lake, and Arbutus Lake Preserves [#45].

The era of railroad tourism began to come to an end before the First World War, by which time the internal combustion engine was drastically changing the way people took their vacations. A decline in railroad map production was symptomatic, as the automobile gradually replaced the train as the principal mode of transportation in the Adirondacks.

After nearly a century of being overshadowed by railroad and inland water conveyances, the common road was restored to its former prominence on maps. Seneca Ray Stoddard was among the first to recognize the significance of the automobile. In 1907, he took the title of his *Map of the Adirondack Wilderness* and changed it to *Road Map of the Adirondack Wilderness*. Three years later, he responded to the demand among touring motorists for a map which would meet their particular needs. The result was *Stoddard's Auto-Roadmap of the Adirondacks, the Champlain Valley and the Hudson River*, published in several editions between 1910 and 1914. The

---

Section of *Stoddard's Auto-Roadmap of the Adirondacks, the Champlain Valley and the Hudson River*, Seneca Ray Stoddard, 1910. The automobile quickly left its imprint on Adirondack tourist maps of the early twentieth century. The photographs printed on this map illustrate accomodations for the visiting motorist, and the captions note where gasoline, oils, auto supplies and repair services were available.

map supplied essential information for the automobile tourist, including locations of garages, auto supply dealers, and gasoline vendors [#47]. The 1910 edition had a rather novel appearance, owing to the inclusion of small photographs depicting hotels, taverns and attractions along the way. Stoddard thus combined his aptitude as a mapmaker with his greater and primary vocation as a photographer.

The automobile required a better system of roads than had horse-drawn vehicles; the state and counties began implementing road construction and improvement projects during the century's first decade. Public agencies issued maps showing the extent of completed and proposed highway improvements. Among these were a 1907 state atlas and county highway maps published as a set in 1912 by the New York State Commission of Highways. In 1916 the commission produced a *Map of the State of New York Showing the Improved Highways and Many Points of Historical Interest*, more clearly aimed at the growing number of car-owning tourists.

Newly-formed auto tour clubs, such as the New York State Automobile Association and the Empire Tours Association, published guides to assist the vacationing motorist. Heir to the earlier railroad and steamboat guidebooks, motorist handbooks served similar purposes by containing maps about routes and conditions of travel, and about attractions and services encountered along the way, such as campgrounds, cabins and motels specially adapted to the needs of the motorist and his car. Many automobile maps covered the entire state, but several were confined to the Adirondacks, among them an *Automobile Road Map of Northern New York Giving the Principal Highways Through the Adirondack and Thousand Island Regions*, published in 1917 by the Santway Photo-Craft Company of Star Lake; also, a map of *Northern New York and the Adirondacks*, published by the United States Survey Company of Rochester, New York in 1925 [#48].

In the twenties, automobile guidebook maps were joined by the type of road map that is widely used today, the folding state map distributed by oil companies. Recognizing the motorist map as a potent medium for advertising their gasoline and oil products, Gulf, Tide Water (Tydol), Sunoco, Texaco and others published maps through cartographic firms like Rand, McNally & Company of Chicago. These were not so much maps as they were folding brochures which contained colorful ads, mileage charts and car maintenance records. The maps themselves noted paved roads, improved roads, dirt roads, numbered state highways, tourist camps, historic and scenic sites, and such features as "Haggerty's Strictly Fireproof Garage," the "Hannah and Henry Motor Car Company," and the "Chazy Landing Auto Ferry." The automobile was changing both the face of the Adirondacks and the maps which depicted the region.

Automobile tourism was a nationwide phenomenon, and though the Adirondacks were shown on state road maps, the area retained its own unique resort character. Tourist agencies formed by local businesses, such as the Adirondack Resorts Association and the Central Adirondack Association, produced maps to guide the motorist within the locality. The users of these maps were unlike those who had consulted regional guidebooks a generation or two earlier: the automobile had democratized the summer vacation by opening the northern woods to middle-class families who ignored resort hotels in favor of cheaper campgrounds, roadside eateries and cottages. Freed of timetables and hotel conventions, the vacationing motorist was constrained only by the availability of decent roads.

Due chiefly to the new vacationer, tourist "attractions" stressing sightseeing rather than recreation prospered on the eastern edge of the Adirondacks, where scenery was combined with historic sites of the Champlain Valley. Historic sites would be joined in later years by scenic drives and amusement parks. Maps reflected this type of boardwalk tourism with such titles as *Historic Shrines in a Summer Paradise* and *A Romance Map of the Northern Gateway* [#49].

Preservationists and Recreationists, 1919-1970

The growth of Adirondack tourism, sparked by nineteenth-century literary travelers, fueled by wealthy vacationers and seasonal residents, and accelerated by the widespread adoption of the automobile, profoundly affected the question of land use. Where conservationists had advocated the management of natural resources for utilitarian purposes, a new generation, today called preservationists, began emphasizing the intangible benefits of aesthetics and spiritual values in an unspoiled wilderness. Tourism was a major contributor to the regional economy, yet the concentration of users in certain parts of the Forest Preserve threatened the environment and wilderness experience that preservationists wanted to protect. Maps of the Adirondack Park reflected ensuing conflicts of overuse, underuse and plain abuse.

In 1919, the New York State Conservation Commission added a new type of map to the resource charts it had been making for years. Published as supplements to the commission's annual reports, the maps *Main Routes of Travel in the Adirondack Mountains of the New York State Forest Preserve* and *Map of Adirondack Canoe Routes* were not aimed at scientists, engineers or bureaucrats, as previous maps had been, but at tourists seeking information about recreation in the Adirondacks [#53]. Other tourist-oriented map circulars followed, including *Map of the Adirondack Mountains and the St. Lawrence Reservation* in 1920, *Map of the Mt. Marcy Region* in 1921 [#54], and *Conservation Map of New York State* in 1930. The

BROKEN GAME LAWS
Less GAME
SPORT
BUSINESS
EVERYBODY LOSES
Your Own Personal Influence Counts
HELP PREVENT VIOLATIONS
COOPERATE WITH THE
CONSERVATION COMMISSION, ALBANY
East Pine Pond
West Pine Pond
Rock Pond
Long Pond
Floodwood
To Ottawa
Wolf Pond
Rollins Pond
Floodwood Pond
Middle Pond
Little Square Pond
Polliwog Pond
Hoel Pond
Green Pond
Little Wolf Pond
Copperas Pond
Little Polliwog Pond
Lead Pond
Whey Pond
Raquette Pond
Tupper Lake Junction
Faust P.O.
Follensby Clear Pond
Square Pond
Tupper Lake
Deer Pond
Fish Creek Ponds
Fish Creek Bay
Simon Pond
Raquette River
Upper Saranac Lake
Weller Pond
Bartlett Carry
Indian Carry
Middle Saranac Lake

Conservation Commission updated and reissued these recreation map circulars in the decades that followed, making sure to advise vacationers to prevent forest fires and heed game laws.

After the First World War, the Conservation Commission assumed a larger role in administering public recreation. It promoted the stocking of fish in streams and established a network of roadside campsites, lean-tos, foot trails, and canoe portages to promote controlled use of the Forest Preserve. Other recreational features included bridle trails, parks, boat launching and public fishing sites, picnic areas, and hunter access points. These were identified on tourist maps such as the *New York State Outdoors Map*, published by the Conservation Department in 1948. Some state-produced recreation maps also noted fire observation towers, game refuges and tree nurseries. Such projects as dam and reservoir building and scientific forestry inspired entire maps during the mid-twentieth century, including charts of Panther Dam, the Sacandaga Reservoir, and the Pack Demonstration Forest [#59].

Outdoor recreation was a major theme of commercially published maps as well. Joining the general motorist maps of the automobile age were specialized maps directed at specific recreational activities and their practitioners. The skiier, horseback rider, angler, hunter, boater, hiker and camper each had his "literature." Among these were: the *Olympic Ski-Trail Map* published by the Adirondack Mountain Club in 1932; the *Cole Bridle Trail Map of Lake Placid* published by the Lake Placid Riding Club in 1940; a *Navigation Map of Cranberry Lake* designed for recreational boaters in 1946; and a 1944 *Map of Blue Mountain, Utowana and Eagle Lakes, Hamilton County, New York*, a combination guide for fishermen and placemat for advertising tourist businesses in the resort community of Blue Mountain Lake.

After the Second World War, theme parks joined historic shrines as prominent features on many tourist maps, including *The Adirondack Picture Map*, published in 1948 by the Adirondack Resorts Press, and *The Scenic Adirondack Lake Shore Routes* [#50]. The success of these attractions revealed that recreation in the Adirondacks did not depend necessarily upon wilderness. Eco-

---

Section of *Map of Adirondack Canoe Routes*, Arthur S. Hopkins & K. F. Williams, 1919. This map represents the entry of the New York State Conservation Commission into the business of making tourist maps. The map not only shows canoe routes (dark lines) and carries (dots), but it also advises the vacationer to respect the resources of the Adirondack Forest Preserve.

nomic needs often were at odds with the preservationist's desire to save the environment.

The conflict between commercial development and environmental stewardship was illustrated in the building and improvement of roads, particularly those encroaching on the state's Forest Preserve lands. The route proposed for the Adirondack Northway, a superhighway connecting Albany and Montreal, called for the taking of several hundred acres of land presumably protected under the "forever wild" clause of the state constitution. A study of the proposal yielded a specialized map entitled *Route "B," N.Y. State P.W. Dept. Champlain Route* in 1957 [#62]. New York's voters approved the construction of the superhighway in an amendment to the state constitution placed on the ballot in 1959.

Land Managers, 1967-1985

In 1967, Laurence Rockefeller proposed the establishment of a federally operated "Adirondack Mountain National Park." The proposal, the subject of a map published that year by the National Survey [#61], never materialized, but the ensuing furor served to focus attention on the common future of public and private lands constituting the Adirondack Park.

Reconciling wilderness preservation and unplanned development led to the creation, in 1968, of the Temporary Study Commission on the Future of the Adirondacks, and the publication, in 1970, of two reports which separately examined state-owned and privately held lands within the Adirondack Park. An accompanying map, titled *Adirondack Park Forest Preserve Classification*, proposed to classify public lands in the Park according to intensity of use. This land-use plan called for a hierarchy of categories ranging from most restrictive to least restrictive: wilderness, primitive areas, wild forest areas, and a blanket category covering campsites, boat launching sites, ski areas, and memorial highways.

One of the commission's recommendations resulted in the establishment of the Adirondack Park Agency (APA) in 1971, and the imposition of state regulation of development on private lands. The agency's *Adirondack Park, Preliminary Private Land Use and*

Section of *Route 'B,' N.Y. State P.W. Dept. Champlain Route*, New York State Public Works Department, 1957. This manuscript map shows three proposed routes for the Adirondack Northway and the encroachment of the westernmost route on Forest Preserve land (the black bars). The map is thus an example of the continuing pressures of human activity on the Adirondack environment.

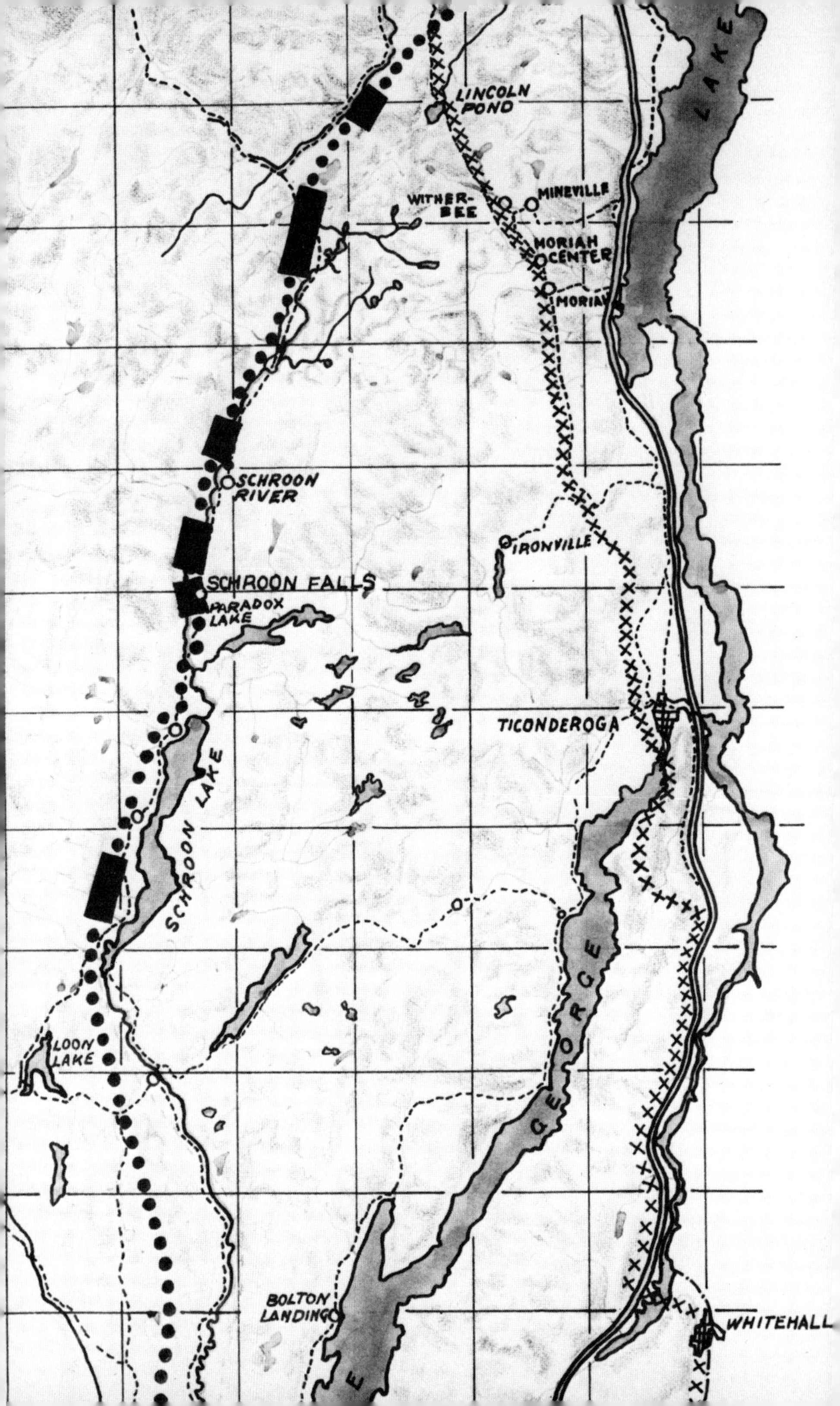

LAKE
LINCOLN POND
WITHER-BEE
MINEVILLE
MORIAH CENTER
MORIA
SCHROON RIVER
IRONVILLE
SCHROON FALLS
PARADOX LAKE
TICONDEROGA
SCHROON LAKE
GEORGE
LOON LAKE
BOLTON LANDING
WHITEHALL

Bloomingdale
ROAD
192
Spitfire Lake
Upper Saint Regis Lake
Saint Regis Pond
Little Clear Pond
Lake Clear
30
50
PENN-CENTRAL RAILROAD
Moose Pond
Lake Colby
McKenzie Pond
Saranac Lake
FOREST HOME ROAD
18
33
Kiwassa Lake
Lake
Saranac
Clear Pond

Development Plan Map of 1972 was the first of a succession of maps, each more finely tuned than its predecessor, which revealed the regulatory tasks that lay ahead for the agency and those affected for good or ill by it [#64]. The map zoned private lands in the Adirondack Park, designating them as areas severally devoted to hamlets, moderate intensity use, low intensity use, rural use, industry, and resource management. In 1975, the APA classified various rivers within the Adirondack Park according to use. Its map, *Adirondack Park Portion of the N.Y.S. Wild, Scenic and Recreational Rivers System*, assigned the designations given in the map's title.

Adirondack towns, aided by professional planners and with the encouragement of the APA itself, did something seldom done before – plan rationally for the future. The map *Ausable Proposed Land Use Plan*, published by the Town of Ausable Planning Board in 1976, is a prime example of local zoning [#65]. It shows the town divided into areas categorized by use – conservation, recreation, agriculture, commerce, industry, and residential.

Since they illustrate a pastiche of human activity, the zoning maps of the 1970s and 1980s represent a final summary chapter in the ongoing history of Adirondack mapping. They present a varied, and often conflicting picture of concerns that have played a major part in the saga of the Adirondacks since the days of Simeon DeWitt and Totten and Crossfield two hundred years ago – the interests of the resident, the local businessman, the industrialist, the farmer, the scientist, the engineer, the tourist and the environmentalist.

Maps have always reflected the various ways in which man has perceived the Adirondacks, from the "waste and unappropriated" land of the colonial explorer to our own generation newly arrived at a full awareness of the values of an unspoiled environment. Maps will continue to reflect changing needs, changing attitudes, and changing technologies. Perhaps a clue about the future of Adirondack mapping is given by a satellite orthophotomap of 1977 titled, *The Adirondacks. . .a portrait from space* [#66]. Cartography and its depiction of man's relationship to the Adirondacks certainly has come a long way since the time of Samuel de Champlain's arrival 377 years ago.

Section of *Adirondack Park Land Use and Development Plan Map*, Adirondack Park Agency, 1974. Recently, maps such as this one have addressed the question of land use control in the Adirondacks. The zoning of both public and private land is shown by the various shades.

Bibliographical Notes

The conclusions presented in the preceding essay are based upon an examination of the map collections of the Adirondack Museum Library. The reader is advised to consult the following secondary sources for more information.

For a background to the history of cartography and mapmaking techniques, the works of Walter W. Ristow, Chief of the Geography and Map Division of the Library of Congress, are highly recommended. Ristow's *Guide to the History of Cartography: An Annotated List of References on the History of Maps and Mapmaking* (Washington, D.C.: Library of Congress, 1973) is the quintessential bibliography. His recently-published *American Maps and Mapmakers: Commercial Cartography in the Nineteenth Century* (Detroit, Mich.: Wayne State University Press, 1985) is the definitive work on the subject and is based upon his earlier publications, particularly the seminal exhibition catalog, *Maps for an Emerging Nation: Commercial Cartography in Nineteenth-Century America* (Washington, D.C.: Library of Congress, 1977). Seymour I. Schwartz and Ralph E. Ehrenberg provide an excellent analysis of mapmaking trends and key cartographers for not only the 1800s, but also periods before and after, in *The Mapping of America* (New York: Harry N. Abrams, Inc., 1980). For the specialized subject of birds-eye views, the recommended source is John R. Hebert's *Panoramic Maps of Cities in the United States and Canada: A Checklist of Maps in the Collection of the Library of Congress, Geography and Map Division* (Washington, D.C.: Library of Congress, 1984). Norman J. W. Thrower's *Maps and Man, An Examination of Cartography in Relation to Culture and Civilization* is a general work which discusses the role of maps in society.

For general histories of the Adirondack region and for references to more specific regional topics and influential figures mentioned in the essay, the *Adirondack Bibliography* (Gabriels, N.Y.: Adirondack Mountain Club, 1958) and *Adirondack Bibliography Supplement, 1956-1965* (Blue Mountain Lake, N.Y.: The Adirondack Museum, 1973) provide a long list of primary and secondary sources, although publications after 1965 are not included. These bibliographies feature sections on noteworthy guidebooks, gazetteers and atlases, and they list some important maps, although not comprehensively or systematically.

Checklist for the Exhibition

The maps in this checklist are arranged in the order they were displayed in the exhibition, "History in the Mapping," held at the Adirondack Museum in 1984 and 1985. The main entry for each map includes the title in bold face, compiler (if known), date, publisher and place of publication in parentheses, overall page size, and scale. Other information about the maps can be found in the notes below each entry. This information includes: known delineators, engravers, lithographers and printers if they are different than the compiler and publisher; publications in which the maps appeared, if applicable; and reproduction information, if applicable. All maps are in the research library at the Adirondack Museum.

Introduction

1. Map of the New York Wilderness and Adirondacks.
William Watson Ely. 1878. (George W. & Charles B. Colton & Co., New York.) 32¼" X 28." Scale: 1" = 4 mi.

The Unknown Wasteland

2. Carte de la nouvelle france. . . .
Samuel de Champlain. 1632. (No publisher noted.) 11¾" X 18" (reproduction size). Scale: not indicated.

This map was represented in the exhibition by a reproduction, titled "Champlain's Map of New France," reduced from the original by David Daughan, engraved by Augustus Tolle, and lithographed by R. H. Pease (Albany) in 1850 for E. B. O'Callaghan's *Documentary History of New York, Vol. 3.*

3. Novi Belgii, Novaeque Angliae nec non partis Virginae Tabula multis in locis emendata.
Nicolao Joannis Visscher II. 1685. (Nicolao Joannis Visscher II, Amsterdam, Holland.) 17⅜" X 22½" (reproduction size). Scale: not indicated.

This map was represented in the exhibition by a reproduction, titled "New Belgium, New England and Virginia," made during the 1970s(?).

4. **Perspective View of the Battle fought near Lake George on the 8th of Sept. 1755, between 2000 English, with 250 Mohawks. . . .**
Samuel Blodget. 1756. (Thomas Jefferys?, London). 12½" X 22¼" (reproduction size). Scale: none for perspective drawing.

Original map engraved by Thomas Johnston (Boston). This map was represented in the exhibition by a reproduction lithographed by Richard H. Pease (Albany) in 1852.

5. **A New and Accurate Map of the present War in North America.**
1757. (Universal Magazine of Knowledge and Pleasure, London.) 11½" X 15¼." Scale: not indicated.

Drawn and engraved by Richard William Seale (London). Map from *Universal Magazine of Knowledge and Pleasure*, May 1757.

6. **A Map of the Middle British Colonies in North America.**
Thomas Pownall. 1776. (John Almon, London.) 20¼" X 34." Scale: 4½" = 150 English miles.

Map from Thomas Pownall's *A Topographical Description of such parts of North America as are contained on the map of the Middle British Colonies, in North America* (London: John Almon, 1776).

7. **The Provinces of New York and New Jersey; with Part of Pennsilvania, and the Province of Quebec.**
Samuel Holland. 1776. (Robert Sayer and John Bennett, London.) 30" X 22½" (size of the northern section of a reproduction). Scale: 1" = 10 mi.

Additions and corrections made to original map by Thomas Pownall. This map was represented in the exhibition by the northern section of a reproduction printed by Headley Brothers Ltd. (Kent, England) and published by Harry Margary (Kent, England) in 1974.

8. **A Plan of the Lands Purchased for the Benefit of Joseph Totten & Stephen Crossfield and their Associates. . . .**
Ebenezer Jessup. 1772. (Manuscript map.) 15¼" X 25¼" (reproduction size). Scale: 1" = 160 chains (2 mi.).

This map was represented in the exhibition by a reproduction printed by J. B. Lyon Company and published by the New York State Engineer and Surveyor's Office (Albany) in 1903 for the book, *Certified Copies of Ancient Field Notes and Maps*.

9. **A Map of the Province of New-York. . . .**
Claude Joseph Sauthier. 1776. (William Faden, London.) 30" X 24." Scale: 2½" = 40 mi.

The Advancing Frontier

10. Map of A Tract of Land in the State of New York called Macomb's Purchase.
Charles C. Brodhead. 1792. (Manuscript map.) 24" X 15⅛" (reproduction size). Scale: not indicated.

This map was represented in the exhibition by a reproduction printed by J. B. Lyon Company and published by the New York State Engineer and Surveyor's Office (Albany) in 1903 for the book, *Certified Copies of Ancient Field Notes and Maps*.

11. Map of the Northern Part of the State of New York.
Amos Lay. 1812. (Peter Maverick?, New York). 30¾" X 49⅞." Scale: 1½" = 10 mi.

Engraved by Peter Maverick (New York).

12. The State of New York with part of the adjacent states.
John H. Eddy. 1818. (James Eastburn & Co., New York.) 38¾" X 44." Scale: 1" = 10 mi.

Engraved by Tanner, Vallance, Kearny & Co. (Philadelphia). Printed by Samuel Maverick (New York).

13. The Tourist's Map of the State of New York.
1827. (William Williams, Utica.) 20⅜" X 29¾." Scale: 1" = 19 mi.

Engraved by V. Balch and S. Stiles (New York).

14. Survey of the Several Routes for a Rail Road from Ogdensburgh to Lake Champlain. . . .
Edward H. Brodhead. 1840. (New York State Assembly, Albany.) 27" X 36¼." Scale: not indicated.

Lithographed by Miller's Lith. (New York). Map from *New York State Assembly Document #43 – A Report of the Commissioners Appointed to cause a survey to be made of the several routes for a rail-road from Ogdensburgh to Lake Champlain* (1841).

15. Untitled [Survey of a Railroad and Steamboat Route from Lake Champlain to the County of Oneida].
Farrand N. Benedict. 1846. (New York State Senate, Albany.) 17" X 62½." Scale: not indicated.

Map from *New York State Senate Document #73 – Memorial of George A. Simmons and six other gentlemen stating the results of a survey of a railroad and steamboat route from Lake Champlain to the County of Oneida* (1846).

16. Map of Long, Forked and Little Forked Lakes from Surveys made July 1874/ Map of Long Lake and Tributaries.
W. B. Cooper, 1874. (Weed, Parsons & Co., Albany.) 19⅞" X 30¾."
Scales: 1" = 4000 ft./ 1" = 2 mi.

Drawn by C. D. Burrus. Map from *Annual Report of the Canal Commissioners of the State of New York* (Albany: Weed, Parsons & Co., 1875).

17. Map of the Rail-Roads of the State of New York.
Sylvanus H. Sweet. 1864. (Weed, Parsons & Co., Albany.) 25" X 30½." Scale: not indicated.

Map from *Annual Report of the State Engineer & Surveyor for the State of New York, and of the Tabulations and Deductions from the Reports of the Railroad Corporations for the Year ending September 30, 1864* (Albany: Weed, Parsons & Co., 1865).

18. Geological Map of the State of New York, by Legislative Authority.
1842. (George E. Sherman & John Calvin Smith, New York.) 36" X 38½." Scale: not indicated.

Printed by S. C. Clark (New York).

19. Geological Map of Clinton County.
Ebenezer Emmons(?). 1842. (W. & A. White & J. Visscher, Albany.) 11⅜" X 8¾." Scale: not indicated.

Lithographed by G. & W. Endicott (New York). Map from Ebenezer Emmons' *Geology of New-York, Part II* (Albany: W. & A. White & J. Visscher, 1842).

20. Agricultural Map of the State of New-York.
Ebenezer Emmons. 1846. (Charles Van Benthuysen & Co., Albany.) 21" X 26⅝." Scale: not indicated.

Lithographed by G. & W. Endicott (New York). Map from Ebenezer Emmons' *Agriculture of New York, Vol. 1* (Albany: C. Van Benthuysen & Co., 1846).

21. Asher & Adams New Topographical Atlas and Gazetteer of New York Meteorological Map.
Lorin Blodget. 1871. (Asher & Adams, New York.) 17½" X 24½." Scale: not indicated.

Electrotyped by the Franklin Type Foundry (Cincinatti). Map from *Asher & Adams' New Topographical Atlas and Gazetteer of New York* (New York: Asher & Adams, 1871).

22. Geological Map of Essex County, New York.
E. V. D'Invilliers. 1879. (Weed, Parsons & Co., Albany.) 20⅛" X 16⅞." Scale: 1" = roughly 3 mi.

Compiled from notes made by Charles E. Hall. Although the map is dated 1879, it is from *Report of the State Geologist for the Year 1884* (Albany: Weed, Parsons & Co., 1885).

23. Map of 1298 80/100 Acres of Lands belonging to the St. Lawrence Mining Co. In the Town of Macomb, St. Lawrence County, New York.
1852. (Arthur & Burnet, New York.) 21¼" X 27⅞." Scale: 1" = 100 ft.

Map from *A Report on the Mines and Lands of the St. Lawrence Mining Co.* (New York: Arthur & Burnet, 1852).

24. Map of a part of the Hochstrasser Lot Dannemora N.Y. . . . .
D. M. Arnold. 1875. (Manuscript map.) 28⅛" X 23⅝." Scale: 1" = 1 chain.

25. Map showing the location of the Pine Timber mentioned in the report of the Surveyors and Lumbermen.
1854. (W. E. & J. Sibell, New York.) 10⅜" X 7½." Scale: 1" = 1 mi.

Map from *The Adirondack Iron and Steel Company, New York* (New York: W. E. & J. Sibell, 1854).

26. A Map made by R. Fairbanks Surveyor for Cheney & Arms. . . .
R. Fairbanks. 1855. (Manuscript map.) 22" X 18." Scale: not indicated.

Tracing of the original map, n.d.

27. A Survey of the Empire Group.
Henry MacNair. 1908. (The Home Educator Company, Detroit.) 43" X 40." Scale: 1" = 10 mi.

28. Map of the County of Essex.
David H. Burr. 1829. (Simeon DeWitt, Albany.) 31" X 21⅝." Scale: 1" = 2½ mi.

Printed by Rawdon, Clark & Co. (Albany) and Rawdon, Wright & Co. (New York). Map from David H. Burr's *Atlas of New York* (Albany: Simeon DeWitt, 1829).

29. Map of the County of Essex.
David H. Burr. 1839. (Stone & Clark, Ithaca, N.Y.) 30¾" X 22¼." Scale: 1" = 2½ mi.

Printed by Rawdon, Clark & Co. (Albany) and Rawdon, Wright & Co. (New York). Map from David H. Burr's *Atlas of New York* (Ithaca, N.Y.: Stone & Clark, 1839).

30. Atlas of Clinton Co. New York.
Frederick W. Beers and assistants. 1869. (F. W. Beers, A. D. Ellis & G. C. Soule, New York.) 16¼" X 13¼" (closed). Scale: various.

Engraved by Worley & Bracher (Philadelphia). Printed by James McGuigan (Philadelphia).

31. County Atlas of Warren N.Y.
Frederick W. Beers. 1876. (Frederick W. Beers & Co., New York.) 16⅛" X 13¼" (closed). Scale: various.

Engraved by L. E. Neuman (New York). Printed by Charles Hart (New York).

32. Post Route Map of the State of New York. . . .
United States Post Office Department. 1914. (United States Post Office Department, Washington, D.C.) 33¾" X 35¾" (northern section). Scale: 1" = 5 mi.

Only northern section of map displayed.

33. New Map of St. Lawrence County, N.Y.
1897. (Edgar G. Blankman, Canton, N.Y.) 72" X 65½." Scale: 1" = 1 mi.

Printed by A. H. Mueller (Philadelphia).

34. Map Showing the Location of the Highway Between Northville and Wells.
C. G. Locke. 1897. (Wynkoop Hallenbeck Crawford Co., Albany.) 22⅛" X 30⅝" (northern section). Scale: 1:10,000.

Only northern section of map displayed.

35. Cram's Railroad & County Map of New York.
1879. (George F. Cram, Chicago.) 17¾" X 22." Scale: 1" = 19 mi.

The Wilderness Resort

36. Secondary Reconnaisance Sketch of Mount Marcy and of the most elevated Lakelet or Pond Sources of the Hudson River.
Verplanck Colvin. 1873. (Weed, Parsons & Co., Albany.) 24½" X 19." Scale: not indicated.

Map from Verplanck Colvin's second *Report on the Topographical Survey of the Adirondack Wilderness of New York* (Albany: Weed,

Parsons & Co., 1874).

37. Sketch Showing the Saranac River Survey (Base Lines) 1878.
Verplanck Colvin. 1878. (Weed, Parsons & Co., Albany.) 18¾'' X 23⅞.'' Scale: 1/126,720.

Map from Verplanck Colvin's seventh *Annual Report on the Topographical Survey of the Adirondack Wilderness of New York* (Albany: Weed, Parsons & Co., 1880.)

38. Lake Placid Sheet.
Henry Gannett. 1894. (United States Geological Survey, Washington, D.C.) 36½'' X 16¼.'' Scale: 1:62,500.

39. Map of the New York Wilderness.
William Watson Ely. 1867. (George W. & Charles B. Colton & Co., New York.) 28'' X 40¼.'' Scale: 1'' = 4 mi.

40. Map of the Adirondack Wilderness.
Seneca Ray Stoddard. 1883. (Charles Van Benthuysen & Sons, Albany.) 32⅜'' X 25¾.'' Scale: 1'' = 4 mi.

Engraved and printed by L. E. Neuman & Co. (New York). Map from Seneca Ray Stoddard's *The Adirondacks Illustrated* (Albany: Charles Van Benthuysen & Sons, 1883).

41. Map of Lake George.
Seneca Ray Stoddard. 1899. (Seneca Ray Stoddard, Glens Falls, N.Y.) 40¼'' X 14¼.'' Scale: 1'' = 1 mi.

Map from Seneca Ray Stoddard's *Lake George* (Glens Falls, N.Y.: Seneca Ray Stoddard, 1899).

42. Map of the Delaware and Hudson Canal Co.'s Railroads and Connections.
Benjamin Clapp Butler. 1879. (Franklin Press, Boston.) 14¾'' X 12¾.'' Scale: not indicated.

Engraved and printed by Rand, Avery & Co. (Boston). Map from Benjamin C. Butler's *The Summer Tourist, descriptive of the Delaware & Hudson Canal Co.'s Railroads, and their Summer Resorts!* (Boston: Franklin Press, 1879).

43. The Health and Pleasure Resorts of New York and New England, the Best Way to Reach Them Via America's Greatest Railroad.
George H. Daniels. 1897. (New York Central & Hudson River Railroad, New York.) 16⅛'' X 39¾.'' Scale: not indicated.

Engraved and printed by The Matthews-Northrup Co. (Buffalo). Map from *Resorts in the Adirondacks, A Region of unsurpassed beauty, A Wilderness and yet a paradise* (New York: New York Central & Hudson River Railroad, 1897).

44. Map of the Chateaugay Railroad and Connections.
1898. (Chateaugay Railroad, Plattsburgh, N.Y.) 15½'' X 18¼.'' Scale: not indicated.

Drawn and engraved by American Bank Note Co. (New York). Map from *A Souvenir* (Plattsburgh, N.Y.: Chateaugay Railroad, 1898).

45. Map of Arbutus Camps, Township 28, Totten & Crossfield's Purchase, Essex County N.Y.
Wesley Barnes. 1899. (Manuscript map.) 30¾'' X 19.'' Scale: 1'' = 50 ft.

Drawn by M. van Mittendorfer(?).

46. Adirondack League Club Preserve.
Augustus D. Shepard, Jr. 1906. (A. R. Ohman Map Co., New York.) 26⅜'' X 21⅜.'' Scale: 1'' = 1 mi.

47. Stoddard's Auto-Roadmap of the Adirondacks, the Champlain Valley and the Hudson River.
Seneca Ray Stoddard. 1910. (Glens Falls Publishing Company, Glens Falls, N.Y.) 32'' X 19.'' Scale: not indicated.

48. Northern New York and the Adirondacks.
1925. (United States Survey Company, Rochester, N.Y.) 26'' X 36⅛.'' Scale: 1'' = 5 mi.

Map from *Northern New York: Adirondacks, Thousand Islands, Lake George & Lake Champlain Nufold Road Guide* (Rochester, N.Y.: United States Survey Company, 1925).

49. A Romance Map of the Northern Gateway.
C. Eleanor Hall. 1934. (No publisher given.) 24⅝'' X 18¾.'' Scale: not indicated.

Original map designed by Carlo Nisita. This map was represented in the exhibition by a reproduction printed by The Holling Press, Inc. (Buffalo) and published by Crown Point Press, Inc. (Elizabethtown, N.Y.) in 1968.

50. The Scenic Adirondack Lake Shore Routes.
Arnold Hutchins. 1957. (Route 9N Association, Inc., Hague, N.Y.) 23¾'' X 18.'' Scale: not indicated.

51. The Saranac Lakes and Fish Creek Campsite Region of the Adirondack Mountains.
R. G. Dustin. 1964. (R. G. Dustin, n.p.) 25" X 38." Scale: 1" = ½ mi.

52. Adirondacks Property of the North Woods Club, Minerva, Essex County, New York.
"C. B. S." 1964. (North Woods Club, Minerva, N.Y.) 19" X 16." Scale: 1" = roughly 2/7 mi.

53. Map of Adirondack Canoe Routes.
Arthur S. Hopkins & K. F. Williams. 1919. (New York State Conservation Commission, Albany.) 14" X 29." Scale: not indicated.

Printed by Redfield-Kendrick-Odell Co. (New York). Map from William G. Howard's *Adirondack Canoe Routes* (Albany: New York State Conservation Commission, 1919).

54. Map of the Mt. Marcy Region.
Arthur S. Hopkins. 1925. (New York State Conservation Commission, Albany.) 17¼" X 26½." Scale: 1" = 1 mi.

Printed by J. B. Lyon Company (Albany). Map from Arthur S. Hopkins' *The Trails to Marcy (Recreation Circular 8)* (Albany: New York State Conservation Commission, 1925).

55. Map of the Great Forest of Northern New York showing boundaries (in red) of the Forest Area and boundaries (in blue) of the Proposed Adirondack Park.
A. Robeson. 1890. (New York State Forest Commission, Albany.) 33" X 29½." Scale: 1" = roughly 4½ mi.

Photolithographed by Julius Bien & Co. (New York). Map from *Annual Report of the Forest Commission, Year ending Dec. 31, 1890* (Albany: New York State Forest Commission, 1891).

56. Map of Water Shed Supplying Reservoirs in Adirondack Forest.
David E. Whitford. 1898. (No publisher given.) 38½" X 28." Scale: 1" = 2½ mi.

Map made to accompany a report made to the State Engineer and Surveyor by David E. Whitford, Nov. 4, 1898.

57. Map of Townships 5, 6, 40 and 41, Totten & Crossfield Purchase, Hamilton County, N.Y.
H. S. Meekham. 1901. (Division of Forestry, United States Department of Agriculture, Washington, D.C.) 22¾" X 20⅝." Scale: 1" = ½ mi.

Printed by J. B. Lyon Company (Albany). Map from *Bulletin #30 of* the Division of Forestry, United States Department of Agriculture.

58. Map of Adirondack Preserve Counties.
1911. (New York State Conservation Commission, Albany.) 26⅛" X 20¾." Scale: not indicated.

Printed by Argus Company (Albany).

59. Pack Demonstration Forest.
Partelow *et al.* Date unknown. (New York State College of Forestry, Syracuse.) 22" X 16⅞." Scale: 1" = roughly 1950 ft.

This item is one sheet consisting of four maps.

60. Oswegatchie River – East Branch, Proposed Newton Falls Reservoir.
R. W. Sherman & E. A. Cullings. 1913. (New York State Conservation Commission, Albany.) 9" X 16⅛." Scale: 1" = 3000 ft.

Printed by J. B. Lyon Company (Albany). Map from *Power Possibilities on the Oswegatchie River* (Albany: New York State Conservation Commission, 1914).

61. Proposed Adirondack Mountain National Park.
1967. (The National Survey, Chester, Vt.) 20¼" X 16¾." Scale: not indicated.

Map from *A Report On A Proposed Adirondack Mountains National Park* (Chester, Vt.: The National Survey, 1967).

62. Route "B" N.Y. State P.W. Dept. Champlain Route.
New York State Public Works Department. 1957. (Manuscript map.) 21¾" X 14¾." Scale: not indicated.

63. Adirondack Park Forest Preserve Classification.
Temporary Study Commission on the Future of the Adirondacks. 1970 (The Adirondack Museum, Blue Mountain Lake, N.Y.) 32½" X 24½." Scale: not indicated.

Map from the Temporary Study Commission's *The Future of the Adirondack Park* (Blue Mountain Lake, N.Y.: The Adirondack Museum, 1971).

64. Adirondack Park Land Use and Development Plan Map.
State of New York Adirondack Park Agency. 1974. (Adirondack Park Agency, Raybrook, N.Y.) 48⅝" X 37¾." Scale: 1" = roughly 3 mi.

Printed by the New York State Office of Planning Services (Albany).

65. Ausable Proposed Land Use Plan.

Town of Ausable Planning Board. 1976. (Town of Ausable Planning Board, Ausable, N.Y.) 17" X 21⅞." Scale: 1" = 8/13 mi.

Map from *A Land Use Plan for Ausable, A Proposal For the Future Development of the Town of Ausable* (Ausable, N.Y.: Town of Ausable Planning Board, 1976).

66. The Adirondacks. . .a portrait from space.

1977. (Land Care, Inc., Boonville, N.Y.) 25" X 18⅞." Scale: 1" = roughly 9 mi.

Based on a NASA Landsat satellite photograph.